EUROPE'S DISAPPEARING WASTE

NEW ANTHROPOLOGIES OF EUROPE: PERSPECTIVES AND PROVOCATIONS

Series Editors:
Michael Herzfeld, *Harvard University*
Melissa L. Caldwell, *UC Santa Cruz*

The anthropology of Europe has dramatically shifted ground from its emergence in descriptive ethnography to the exploration of innovative theoretical and methodological approaches today. This well-established series, relaunched by Berghahn Books with a new subtitle, invites proposals that speak to contemporary social and cultural theory through innovative ethnography and vivid description. Topics range from migration, human rights and humanitarianism to historical, visual and material anthropology to the neoliberal and audit-culture politics of Schengen and the European Union.

Recent titles:

Volume 17
Europe's Disappearing Waste: Unanticipated Conflicts between Removal and Greening
Daniel Sosna

Volume 16
The Jewish Maghreb: North African Experiences in Greater Paris since 1981
Samuel Sami Everett

Volume 15
A Border Island on the Crossroads of History: Lampedusa and the Mediterranean
Dionigi Albera

Volume 14
Where Saints Show Respect: Mafia, Modernity, and Rituals of Power
Berardino Palumbo

Volume 13
Lurking Cold War: Life Through Historical Communion
Stavroula Pipyrou

Volume 12
The Spectral Within: Fascism, New Towns and their Contemporary Lives
Elena Miltiadis

Volume 11
Rethinking a Radical Reputation: An Athens Neighbourhood between Media Hype and Social Experience
Maria Kenti Kranidioti

Volume 10
Haunting Futures: Crisis, Migration and Anticipation in Iceland
Marek Pawlak

Volume 9
Ritual, Rapture and Rebellion: The Making of Market, Mercy and Meaning Amongst the Gitanos of El Rastro
Marianne Blom Brodersen

Volume 8
Tactical Citizenships: Encounters with Everyday State in the Republic of Cyprus
Theodoros Kouros

For a full volume listing, please see the series page on our website: https://www.berghahnbooks.com/series/new-anthropologies-of-europe

EUROPE'S DISAPPEARING WASTE

Unanticipated Conflicts between Removal and Greening

Daniel Sosna

berghahn
NEW YORK • OXFORD
www.berghahnbooks.com

First published in 2026 by
Berghahn Books
www.berghahnbooks.com

Library of Congress Cataloging-in-Publication Data

A C.I.P. cataloging record is available from the Library of Congress
Library of Congress Cataloging in Publication Control Number: 2026018817

British Library Cataloguing in Publication Data

A catalogue record for this book is available from the British Library

EU GPSR Authorized Representative
LOGOS EUROPE, 9 rue Nicolas Poussin, 17000, LA ROCHELLE, France
Email: Contact@logoseurope.eu

ISBN 978-1-83695-551-1 hardback
ISBN 978-1-83695-552-8 epub
ISBN 978-1-83695-553-5 web pdf

https://doi.org/10.3167/9781836955511

For Lojza

CONTENTS

FIGURES AND TABLE

FIGURES

TABLE

PREFACE

I still remember a coffee break during a workshop on moral economy in Oxford, where one of the participants approached me, saying, "You really zoomed in!" and raising her eyebrows. It was a friendly reaction to my paper after I expressed admiration for her narrative about large-scale processes that she masterfully described in her paper. But there was another layer of meaning, I believe, and it was not the first time I heard something similar. In times when academic debates feature planetary crises, global flows, and all the associated theorizing, it may seem like an anachronism to describe the complexities in the reasoning of a few interacting landfill workers. This approach may lose another layer of charisma when it comes from a country such as Czechia, where the grandeur of colonial empires and the planetary priorities of humankind may, as some believe, dissolve into local-mindedness. I can imagine that my ethnography may be viewed as too particular and descriptive for some readers. It is intentional because it reflects what I consider the core of anthropological knowledge production: an opportunity to relate to others and imagine what their lives may taste like vis-à-vis larger societal processes. Moreover, the ability to describe in detail various actors, events, and places is an essential prerequisite for building any theoretical argument that scales up and reaches higher levels of abstraction. Such knowledge may resist the tooth of time and changes in theoretical fashions.

Shedding light on the lives of specific persons in particular places is also a way to repay my interlocutors, who allowed me to spend time with them and learn about their lives and the ways they perceive the world. It is impossible to name all of them, and I can only use the first names or nicknames of those who are unlikely to be pinpointed, given my promise to keep their identities hidden, as most wished. Still, I would like to thank at least the most crucial people for my research. I am grateful to Míra, Pavel, Pepa, and Roman for allowing me to work with them, sharing their stories with me, and teaching me how to scavenge and repair discarded items. I appreciate the trust of Bakr, Michal, Milan, Pavel, and Tibor, who enabled me to join their waste-picking activities and learn how to disassemble discarded devices with basic tools. The discussions with Pavel, which were marked by mutual curiosity and openness, were especially valuable to me. Additionally, I would like to thank the managers who permitted me to conduct research at their landfills and who shared a wealth of helpful information regarding the rules, practices, and challenges in waste management with me. Josef and Roman were very kind in taking me to their incinerators multiple times, and they shared their perspectives on the practices with me. Also, I am very grateful to an anonymous landfill headman who was brave enough to reveal information about their 'data cooking' and reporting to the state. Without Ondra and Petr, I would know much less about the operation of the state administration, political lobbying in the waste industry, and the environmental history of places used for waste disposal. My former student, Anna, provided me with valuable insights into social work with families living in social exclusion.

I believe that a sense of description of the particular and narrativity is something that I share with my dear friend and colleague, David Henig. We have shaped our visions of anthropology during the last two decades through mutual discussion. While my four-field training and experience from the United States initially allowed me to share my views on anthropology and academia with David, our roles have evolved over time. As David has become a respected social anthropologist, he has brought his field experience, epistemic tuning, and visions of anthropology to the forefront. I am deeply indebted to him for facilitating a transition from my previous archaeological orientation to the ethnographic one.

I highly appreciate all the discussions and research collaborations with my former and contemporary doctoral students Lenka Brunclíková, Michaela Jaterková, Pavel Mašek, and Bára Stehlíková. Working on projects together meant that we shared a lot, read the same literature and each other's writings, and discussed our experiences, potential interpretations, and theoretical inspirations. Clearly, we influenced each other's thinking substantially. I do my best to acknowledge their input whenever it is relevant. If I omit anything, please bear with me.

I can hardly capture all the support, collegiality, and enthusiasm from my home Department of Ecological Anthropology in Prague. Our department has been dynamic and inspiring, and I am proud to be part of it. Over the last seven years, I received various kinds of support and feedback from Luděk Brož, Giovanna Capponi, Thorsten Gieser, Petr Jehlička, Paul Keil, Ognjen Kojanić, Laura Kuen, Gary Marvin, Maike Melles, Kieran O'Mahony, Annika Pohl Harrisson, Marianna Szczygielska, and André Thiemann. Other colleagues from our Institute of Ethnology, such as Nikola Balaš, Pavel

Horák, Jan Ort, Michal Pavlásek, Stephanie Rudwick, and Michal Šípoš, have directed me to new literature or carefully listened to and critiqued my ideas. Special thanks go to Martin Fotta, whose interest in economic anthropology overlapped with mine, and his comments on the book proposal and Chapter 4 were invaluable. Our director, Jiří Woitsch, deserves recognition for connecting me with new scholars and for his support during the pandemic.

Multiple scholars shaped my pathway to this book in various ways. Roman Figura was a great collaborator on ravens, bringing the perspective of an ornithologist. We spent a few days in the field together and gave collaborative presentations. Catherine Alexander, Miguel Alexiades, David Edwards, Anna Kacperczyk, Michal Lehečka, Patrick O'Hare, Karolína Pauknerová, Dagna Rams, Tony Robben, Andrew Sanchez, Tereza Stöckelová, Alison Stowell, Jason Waite, Jeremy Walton, Rick Wilk, Markéta Zandlová, Patrik Zapata, and Amy Zhang became familiar with my academic texts, proposals, applications, or presentations and helped me in different ways. I am grateful to Maan Barua, Dimitra Kofti, Nicolette Makovicky, and Jörg Wiegratz for inviting me to their workshops, where I had the opportunity to discuss my ideas. The same holds true for Chris Hann, whose multiple kinds of support have been helpful for me and my students. Although I left my previous job at the University of West Bohemia with some bitterness, I must acknowledge Petr Lozoviuk, the departmental chair, for supporting the field research that enabled me to explore life at the landfills. Jan Váně deserves acknowledgment for his support during my transition to Prague. I quit my job at the university without having a new one. Jan offered me a temporary job to help me get over this challenging time. He knew well what it meant to have two kids at home. Jan Mareš's critique was helpful in rethinking my own academic and political biases and the potential of my work to impact the 'real world' outside academic theorizing. Kateřina Joštová deserves acknowledgement for her administrative support.

Earlier versions of Chapters 2, 3, and 5 appeared as, "Saving and Wasting: The Paradox of Thrift in a Czech Landfill," in *Thrift and Its Paradoxes: From Domestic to Political Economy*, edited by Catherine Alexander and Daniel Sosna, Berghahn (2022): 162–84; "The Inner Dynamics of Moral Economies: The Case of Waste Management," *East European Politics and Societies* 38(1) (2024): 321–38; "Circular Economy of Wastewater: Recirculation, Spinning, and Rolling to the Future," in *Circular Economies in an Unequal World: Waste, Renewal, and the Effects of Global Circularity*, edited by Patrick O'Hare and Dagna Rams, Bloomsbury, 153–71; "Magnetism of Strangeness: Silenced Histories of Landscapes," *History and Anthropology* 34(3) (2023): 390–408.

I want to thank Tom Bonnington from Berghahn, and series editors Michael Herzfeld and Melissa Caldwell, for support, encouragement, and inspiring ideas. The insightful suggestions and careful reading of the manuscript offered in the anonymous review were invaluable. Berghahn production department helped me to finalize the manuscript, especially Ivan Butler who copyedited the manuscript. I must acknowledge Patty Gray's engagement with my earlier writing projects, which forced me to think more about my writing. Shelby Brewster was so nice to prepare the Index for this book.

I would like to thank my wife, Veronika, and my two daughters, Aneta and Elen, for their patience in allowing me to conduct the field research, which was challenging for us in many ways. The landfill odor is not easy to remove. Fortunately, I lost my hair over the years! I did not have to face the trouble that the landfill worker Petr, with his long mullet, described in this book. Still, my dirty and bad-smelling trousers, jackets, and muddy boots were often around the house, and not everybody was happy. My commitment to follow the people who had to deal with garbage regardless of the degree of disgust, toxicity, or contagion meant that I also exposed my family to some of those dangers. This required a lot of understanding from my relatives. Additionally, I am grateful for all the challenging questions and for being pushed to think about my research in non-academic ways.

This book would not have been possible without the support of the Czech Science Foundation under grant GA20-06759S. It would be an oversimplification to acknowledge only this major support. Ethnographic monographs need time and patience. My previous project, which was funded by the same agency (GA14-03314S), enabled me to make the initial phase of my research. The *Praemium Academiae* (AP2303) grant supported the last writing phase. A position at the Academy of Sciences enabled me to continue writing during periods without research funding. I am grateful to Petr Jehlička and Luděk Brož, who understood that I had to complete this writing project, despite other duties and commitments emerging.

INTRODUCTION

> No matter ever vanishes. It can, at most, change its form. Rubbish is immortal, it pervades the air, swells up in water, dissolves, rots, disintegrates, changes into gas, into smoke, into soot, it travels across the world and gradually engulfs it.
>
> —Ivan Klíma, *Love and Garbage* 1991[1988]: 8

It is September 2021. A truck with a covered body is backing up to the edge of the Smoke landfill.[1] We stand at both sides behind the car and show the driver when to stop to get as close to the edge as possible. When the driver stops and gets off the truck, he takes on thick leather gloves and greets us with a brief and hardly recognizable mumbling. We respond similarly. The eyes of my fellow waste pickers are full of excitement. When the guy pulls the stick to open the door, everybody wonders how much copper might be inside. No, it does not look good. A mass of sandpaper cuts along with paper and plastic strips starts falling out. When the driver tips the entire load, waste pickers briefly walk around the pile, lift and drop a few large pieces of paper, and walk away with disappointment. No copper cables this time, just a lot of sandpaper.

While my friends look uninterested, I look with astonishment at a pile reaching up to my shoulders and taking space of about six meters in diameter. It consists mostly of diverse sandpaper: coarse, fine, red, green, rectangular individual sheets as well as heavy rolls that must have several meters of the sandpaper still neatly rolled; I cannot even move them on my own. They are too heavy. Wait a minute. The sandpaper smells fresh. It is intact! I am perplexed to see such wastefulness, while recalling how expensive even tiny sheets of sandpaper are in the store.[2] My buddy Fero explains: "I heard that they have new machines for grinding and do not need this kind of sandpaper anymore. Take some. I already have plenty of it at home." One of the local companies producing parts for the automotive industry decided to send the truck with their unwanted sandpaper to the Smoke landfill, hoping it would be an easy way to make it disappear. Who would care if those hundreds of kilograms of unused sandpaper were dumped? For the rest of the day, I contemplate this approach to resources as well as the landfill's capacity to make such disappearance possible. A few incoming drivers who take sheets from the pile indicate that disappearance via dumping is limited. The material comes back.

Three days later, I watch the evening news on TV. A host brings dramatic footage of a fire accident at a landfill, depicting dark gray fumes emitting from the burning garbage. A subsequent visual shift to a new incinerator with a slick design and modular plasma-based facility embedded in a transport container provides examples that contrast the terrifying heaps of garbage disposed of at the landfill. My younger daughter makes an ironic comment: "Gosh, your waste again!" One can tell that she is not excited about her dad's work, but her comment is significant because of the word *again*. Yes, again. To give another spin to Joshua Reno's (2016a: 5, 227) phrase that "garbage keeps coming," my daughter's note suggests that garbage keeps coming despite being physically disposed of: it tends to reappear as a sign with surprising pervasiveness. It inhabits news, popular magazines, discussions on the radio, adventures of Peppa Pig, criminal movies and books, and, of course, scholarly studies. While I consider my professional bias of seeing waste everywhere, its pervasive semiotic presence in public space is beyond doubt. Interestingly, garbage and landfills seem to be a powerful representational couple tied with an invisible umbilical cord. They appear in public spaces to convey divergent messages: wastefulness of contemporary capitalism, climate crisis, environmental pollution, the necessity of market solutions in waste management, technological hopes, or crime. It is striking how much such representations flourish, given the widely shared conviction to keep the waste matter away or let it disappear (see Alexander 2016; Alexander and O'Hare 2023; Reno 2016a).

The co-occurrence of garbage and landfill signals a specific kind of relationship. It can be understood as a hyphen (Fine 1994; Ingram 2024). I am interested here in hyphen beyond its mere linguistic dimension. It can be approached as a tool for exploring connectivity (Ingram 2024: 366). When I talk about the invisible umbilical cord above, I refer to a relationship that can be detected analytically without being acknowledged in everyday practice. It goes without saying that garbage evokes landfill and landfill evokes garbage. When one attempts to get rid of garbage, it may entail getting rid of landfilling as its semiotically most pervasive twin. Similarly, when one attempts to get rid of landfills, it is also a way of getting rid of garbage.

As Robyn Zink and Lisette Burrows say, the hyphen "denotes a relationship at work" (2008: 262). Although I would not go so far as to suggest that the hyphen embodies a mutually constitutive kind of entanglement (Ingram 2024: 366), the relationship between garbage and landfill, or dump, is intimate and forged by centuries of disposal practices.

Resistance of garbage to disappearance is not only a matter of representation in media and public discourses. Diverse kinds of waste tend to resist attempts to be contained or enclosed (Reno 2016b; Resnick 2024b; Zhang 2020). Waste crosses boundaries. It flies over fences, spills over containers, leaks through liners, and moves between categories. Its unruly spread appropriates space via pollution (Liboiron 2021; Serres 2011) and exposes diverse forms of life to the effects of slow violence (Doherty 2019; Nixon 2011). As Ivan Klíma argued in the initial quotation, waste resists disappearance through its openness to transformation (see also Stamatopoulou-Robbins 2020: 23). Waste may become a ghost that troubles the narratives of progress as it haunts us (Gan et al. 2017). There are, however, transformations that are not necessarily negative and harmful. They rely on creativity and skill mobilized to generate satisfaction from waste labor (Sanchez 2020), tinkering (Reno 2016a), or repair (Martínez and Laviolette 2019). These transformations may even generate a sense of reliability in a world full of precarity (O'Hare 2018) and enable garbage to reappear as life-giving rather than only abject and life-threatening (Millar 2018: 32). The reappearance of waste simply has multiple pathways and modalities.

This book is about human aspiration to make waste disappear and the impossibility of achieving it. While the original meaning of *appearance* is intimately tied to visuality, vernacular understandings of disappearance tend to embrace a wider field of stimuli, which relate to other senses and cognitive processes. This is the wider meaning of the verb *zmizet* (to disappear) in Czech. The aspiration to make waste, or garbage in particular, disappear goes way beyond the visual sense. It should prevent waste from troubling us in any imaginable way: being seen, smelled, felt on the skin, tasted, or haunting our imagination. This conceptualization of disappearance covers the whole spectrum of relations to entities that vanish. One should not even have an idea popping into one's mind when disappearance as a project succeeds. This kind of aspiration is ubiquitous in Czechia, and traverses scales from a single disposal of a spoiled ham to large accumulations of garbage in Europe. They all should disappear.

I delve into the world of those who come into direct contact with garbage and its unruly offshoots to shape what stays, becomes transformed, or is left to rot. Via immersion into the lives of landfill workers, waste pickers, drivers, and managers, their relations to material flows, other organisms, societal expectations, and valuations, I trace how these lives unfold within the major project of disappearance, while simultaneously altering or resisting the goal of the project itself. Building upon what I have experienced over the years of my research on Czech waste streams and infrastructures, I argue for a need to reconsider the ubiquitous conviction that waste must disappear (see also Reno 2016a: 217). This reflexive exercise requires a critical rethinking of two associated ideas that also tend to be taken for granted. The first is a belief that the ideal of disappearance can be best achieved via technological innovation or an economic model based on circulation that creates a sense of infinity. I will demonstrate that neither technological innovations nor a turn to economic circularity can resolve the problem of how to relate to waste. The second idea is a complete outsourcing of waste handling and treatment by the forces of the market. In contrast, I propose leaving space for non-market mechanisms, which would prevent the world of discards from being completely swallowed by the hungry trap of the market.

This book's theoretical aspiration is twofold. First, I unpack various regimes of disappearance and reappearance to capture their multifaceted quality and the pervasiveness of their operation. Such a hyphenation of disappearance and reappearance challenges a simplified view that the organisms' relationship to waste can be understood via the lens of presence and absence. In contrast, disappearance and reappearance are contingent upon the eye of the beholder and the regime at work. Second, I propose to think with fluidity (Strang 2014), whose conceptualization goes beyond the material properties of waste. Taking the already existing term 'waste stream' as a starting point, I embark on a journey exploring diverse flows within these waste streams (cf. Hird 2021). The metaphor of fluidity enables me to build relations between the materiality of waste, associated social practices, and discourses dissolved in public space. Fluidity enables me to capture underflows, backflows, or sedimentations that can serve as alternatives for thinking about informality. Moreover, I use fluidity as an epistemic tool to see what kind of knowledge emerges when a researcher steps into the muddy waters of waste streams and lets them set the direction of floating.

DISAPPEARANCE AND REAPPEARANCE

In August 2012, I examined a sample of household waste from one of the urban quarters of a nearby city along with Lenka and Viktor, two of my students, who joined me in our small garbage project. Our tent was built on the edge of the active zone of the Pureland landfill. It was the second week of our field research, and we still struggled a bit with the annoying stench from a nearby chimney designed to release landfill gas. The last several days were sweltering, so we were all looking forward to the end of the day. At about

4 pm, when the landfill's surface reached its peak temperature, and the air shimmered, we suddenly felt a wind coming from nowhere. Ten meters from us, a spinning air cone started moving across the landfill while sucking all the objects that were light enough to be lifted and became part of what looked like a composite figure twisting in a crazy dance. We were astonished at the amount of plastic and paper material ascending and moving to the northeast. Czechia was not a country where one could experience tornadoes, so even this dust devil was unusual.[3] While most members of this strange flying squad fell to the ground or got caught in treetops near the landfill, a plastic bag decided to enjoy flying and embarked on a journey. We stopped working and watched "the parachute" rise and rise. It kept moving northeast toward the city until it became a tiny spot in the sky. It did not seem to descend. I could not help but think of Jules Verne's novel *Five Weeks in a Balloon*, while imagining the worlds the bag might see if it takes advantage of the unusual weather conditions. After half an hour of watching, we started speculating on how far it could get—a few kilometers for sure. The main point, however, was the ability of this tiny adventurer to resist disappearance and make itself a hero of the day visible not only to us but to an unknown number of other creatures who could speculate about its origins and destiny.

This vignette describes one of the senses of disappearance and reappearance. The plastic bag, which was supposed to be buried at the landfill, took a different trajectory. This book will explore the dialectics of disappearance and reappearance through three interconnected registers. Although it is not entirely possible to separate these registers because of their overlaps and mutual interdependence, it is useful, analytically speaking, to classify different ways of thinking with disappearance. It helps to clarify the range of processes where disappearance and reappearance operate.

The first register focuses on the disposal of unwanted matter, which is pushed out of sight but comes back via acts of recovery, repair, exchange, leakages, or unexpected travels facilitated by both human and nonhuman actors. It includes tactics of mixing different things to conceal their properties or origins. It refers primarily to the movement or flow of matter and transformations that enable former waste to reappear. This register is inherently tied to the materiality of waste with all its contingencies and repercussions.

The second register is semiotic and processual. Garbage may index a family's thriftiness that prevents careless use of resources and the presence of polluting forms of scrap metal processing or borders. Garbage may symbolize impurity or unruliness. It may get contained, enclosed, or transformed to reduce its semiotic capacity to signify. Alternatively, it can be mobilized and rendered present as a sign around waste infrastructures, in media, or in public talks to convey various meanings in service of specific goals or generate unintended consequences. Importantly, many signs do not refer to garbage as an object but rather to practices that engage with garbage in multiple ways.

The third register relates to the disappearance and reappearance of landfilling as a technology with all its relations. During the last several decades, the star of landfilling has been fading out in the European Union (EU) and slowly being replaced by other technologies for garbage management. This kind of disappearance, however, contrasts with the attention landfilling has received as a synecdoche of all wastes that plays a vital role in the social mechanism of giving up as a solution to the crisis.

The concept of fluidity connects these registers. It feels strikingly familiar to the world of sanitary engineering with its concept of *waste streams* that serves as a tool to "imagine flows of different materials that have distinct properties and are headed for different destinations" (Reno 2015: 559). Although theorizing about fluidity often starts with water (Krause and Strang 2016; Strang 2014), its goal goes beyond materiality of fluids. It is an invitation to think with fluidity as a metaphor of continuous change that refers to molecules, barriers, relations, negotiations, and experience alike. What is particularly useful for my exploration of fluidity of matter and practices, is Zygmunt Baumann's succinct observation that, unlike solids, fluids "are not easily stopped" (2000: 2). This resonates with the movement of wastewater and practices of waste pickers described in this book. Even more importantly, fluidity enables specific forms of coexistence. As Annemarie Mol (2002: 143) argues, fluidity has a capacity for inclusion that enables coexistence of what is markedly different. This is relevant for imagining practices of different actors who shape the waste streams.

Thinking with flows is a way to capture the unruly dynamic of waste movement in space. Myra Hird (2021) uses flow as a conceptual tool to approach the journeys of Canada's waste through regions and countries. Also, Hird (2021: 178) uses flow to highlight waste's capacity to challenge social categories and leak through. Waste, in other words, is hard to immobilize because of its proclivity to escape and flow further through physical and social forms of containment.

Although some flows can be discerned and sensed easily, others remain hidden. Cleo Wölfle Hazard (2022) developed the concept of underflows to capture dynamism in hydro-ecology and extended it as a conceptual tool for thinking about diverse unseen flows. Hazard's metaphorical thinking moves from the river water that seeps through gravel and soil to form underflows beneath the surface to the spheres of discourses and politics. Underflows serve to direct attention to "latent discourses that lie hidden or disregarded, like fluid underflows are hidden beneath a stream's bed" (Wölfle Hazard 2022: 8). To follow this kind of thinking, I explore a gamut of fluidity from the everyday underflows of waste matter in the tipping zones of landfills to the underflows of ideas about rescuing value. There are two analytical moments that deserve attention. First, the flows do not exist only on the surface where discards move through the conduits of waste infrastructures. There are invisible underflows accessible only to those who have intimate knowledge and experience with them. Second, matter does not have to be the only entity flowing in the waste streams. There are flows of ideas, signs, discourses, pollution, or harm that do not make it to the surface and remain hidden.

* * *

Disappearance is an existential quality of life. Organisms grow and mature, relations with friends are formed and last, but eventually all disappear. This sense of disappearance is closely tied to death. The entropic power of the universe always wins, sooner or later. Humans have attempted to overcome the existential weight of disappearance. Potions for eternal life, incantations, deals with the devil, and medical interventions all try to prevent the disappearance caused by death. One might even try to outsmart death, as Ingmar Bergman's figure of the knight Antonius Bock tried when he challenged death with a chess game. All these attempts have been futile so far.

The end of life, however, represents a social arena with noticeable significance. From mortuary studies, we know that dying and death are not just social phenomena related to loss and sorrow but also an opportunity for the living (Metcalf and Huntington 1991; Nešporová 2021; Robben 2004a). Mortuary rituals are arenas where the living meet and negotiate with each other. Ironically, mortuary rituals are often performed for the living (see Gillespie 2001: 78; Parker Pearson 1982: 112; Verdery 1999), although native conceptualizations of soul-doubles and their mutual relationship with the souls of nonhuman beings make the simplified contrast between the living and the dead more complex (Brož 2024). Ritual practices associated with death and ideas about dying can be mobilized in the service of various ends (Lock 1996), including negotiations of group belonging and distantiation from others (Fotta 2019). Since the death of organisms and things generates a need for subsequent management of their absence, the process invites similar engagements (Hetherington 2004: 168). Sometimes, things and human bodies can even disappear together at dumps as "society's discards" (Millar 2018: 62). All this suggests that it might be productive to approach disappearance at landfills as an arena that creates opportunities known from the contexts when human life ends (cf. Reno 2016a: 218).

There are ways to cope with the inevitability of disappearance, thought of as death. Disappearance can only be a matter of perspective. Cross-culturally shared ideas about death and regeneration of life, analyzed by Maurice Bloch and Jonathan Parry (1982), embody the transformative solution to the problem of vanishing life. The social rupture caused by death is solved via the embracement of the belief that the essence of life does not disappear entirely; it only transforms.[4] Rebirth negates the finality of death and the disappearance of life. Building upon Robert Hertz (1960) and his vision of mortuary practices as anomy-avoiding mechanisms, Bloch and Parry (1982: 15) demonstrate that the representation of death and rebirth as part of a cycle has consequences for social resilience and order-making. Three decades later, David Graeber (2012) extends the concept of cycling and transformation into the economy to argue that the ideas about death and the regeneration of life, on the one hand, and recycling, on the other, have something in common. They both solve the problem of keeping the status quo while being exposed to omnipresent change. Neither organisms nor things vanish; they only undergo transformations imagined as cycles.

Disappearance acquires a polysemic quality. The concept may refer not only to the state of being dead but also to vanishing from being sensed, ceasing to be known, or having an effect. Different senses of disappearance show diverse ways the process unfolds. The disappearance of the Ndembu initiates via their seclusion in the forest lodge during the circumcision ritual serves as a vehicle for ritual transition from one state of a human being to another one (Turner 1967). This sense of disappearance is always temporary, and vanishing from sight carries the promise that the entity will return. In contrast, Antonius Robben's (2004b; 2018) work on terror in Argentina depicts disappearance as a repressive tactic of the state's armed forces that abduct their political opponents and make them disappear for good. This kind of disappearance creates uncertainty among the opponents and prevents revenge or at least a proper funeral. Only the absent dead bodies that would never appear again serve the goals of what Jason De León (2015: 71) calls *necroviolence*. This sense of disappearance does not carry the promise that the person will come back. In contrast, uncertainty and doubts about the future serve the political goals of those in power; they destabilize Argentinian guerillas or contribute to Prevention Through Deterrence among De León's migrants on the US-Mexico border. What remains after such disappearances are memories and signs such as personal belongings, photos, and texts that can trigger these memories.

In waste and discard studies, the disappearance has received particular attention. Martin Melosi's (2005; 2020) historical research demonstrates that spatial solutions in waste management following the logic "out of sight, out of mind" can be a powerful way to make waste disappear. This process can be thought of as "placing absences" (Hetherington 2004: 159). In Hetherington's view, the placement of absences is an essential part of social life and an important vehicle for the performance of social relations. Absence is as socially constitutive as presence, and its significance goes way beyond the themes of waste and discard (Bille, Hastrup and Flohr Sørenson 2010). Discard, nonetheless, is an arena where the management of absence has far-reaching consequences for thinking about disposability, socioeconomic relations, or understanding self and the ways we inhabit the world (Reno 2016a). Spatial management of waste and its disappearance does not mean that one has to make a shortcut claiming that waste is necessarily matter out of place (Liboiron 2022). As Liboiron points out, waste is rarely out of place in a sense proposed by Mary Douglas (2002). Waste is often in place and its spatial management reflects power relations in society (see also Liboiron and Lepawsky 2022).

Keeping waste inaccessible to senses can be understood as an act of hiding. Living in societies where Durkheim's organic solidarity dominates inevitably results in reliance on others, which also means that specific processes and relations are hidden to everyday workings of social life. Such an absence can be taken to the extreme of the modernist insistence on self-creation, which denies the role of others in this creative process (Graeber 2007: 100). Building upon Graeber's ideas, Reno (2016a: 14) argues that landfills intensify reliance on others while simultaneously hiding this reliance. I suggest that attention to the effects of hiding can move this line of reasoning further. As David Berliner (2022) shows, one of the effects of hiding approached from the perspective of the "gymnastics of the self" is the removal of constraints on action. Berliner describes the effect of a mask in promoting action that would not happen without the mask hiding a person's identity. Mask provides a barrier. It constrains but simultaneously opens possibilities. Imagining society as a person and hiding tactics in landfills' placement (remoteness, environment) as a kind of mask, we might consider what such hiding does. I argue that keeping landfills with garbage out of sight enables not only the perpetuation of specific, and not necessarily sustainable, ways of life but, more importantly, also removes constraints on actions that literally get off the leash. Dumping already sorted plastics ready for recycling, tombstones with the names of the deceased, intact packed food, or perfectly usable furniture can happen only because the disposal is hidden and the discards are supposed to disappear. Without the "mask," it would be much harder to perform not only such a disposal but also enable the degree of consumption that generates excess. This view, I believe, resonates with Francisco Martínez's (2025: 6) claim that opacity, secrecy, and hiding are not marginal effects emerging in a modern society, but they are intrinsic to it. Martínez, however, does not envision hiding as a practice that would simply perpetuate the *status quo*. He elaborates on the idea that hideouts are spaces for creative opportunities to transform the hidden and ambiguous into something else and potentially challenge the established configurations of power. In this perspective, the act of separation simultaneously carries the potential to establish relations and to reengage with the world across time and space.

The disappearance of waste has a strong temporal dimension. Interestingly, it is not a simple linear process based on the capacity for decomposition. As Romain Garcier (2012) demonstrated in his study of nuclear waste treatment, a circular movement between storage facilities can create a sense of disappearance. Viewed from a single storage location perspective, nuclear waste may be gone, but the temporal game based on circulation only postpones the final solution. The disappearance of waste is situated within multiple coexisting temporalities (Reno 2016b; Weber 2022b), which can be conceptualized as heterochronic (Sosna 2017) or chronotopic (Alexander 2023).

Catherine Alexander and Patrick O'Hare (2023) provided a key intervention in theorizing about the "disappearance of waste from view." Putting knowledge into the center of their attention, they developed a classification of the "technologies of unknowing," which are supposed to deflect attention away from wastes. Their elaborate classification includes temporal, spatial, epistemological, calculative, and rhetorical technologies. In addition to the spatial and temporal techniques mentioned in the previous paragraph, Alexander and O'Hare describe a range of techniques based on silencing or development of the regimes of secrecy, changing modes of measurement, calculation, scalar shifts, as well as ways waste is renamed or relabeled to become part of discursive games with words. Although many of these techniques are based on intentional decisions to make waste disappear, less conscious forms of ignorance are at play as well. As Judith Bovensiepen and Mathijs Pelkmans (2020) demonstrate, "wilful blindness" includes both conscious and unconscious forms of ignorance. The wide coverage of various modalities of unknowing or not knowing provides a productive conceptual framework for analyzing the disappearance of garbage, landfill leachate, or landfills as such, which will be discussed in the following chapters of this book.

Not knowing refers not only to the absence of knowledge but it may also signify more. As Milan Kundera suggests in his *Unbearable Lightness of Being*, people design facilities to make their bodily wastes disappear in the darkness of "invisible Venices of shit." Kundera's elaboration on the conduits facilitating the disappearance of this particular kind of unwanted matter includes ideas beyond the notions of visibility and knowing.[5] In addition to the recognition of a waste's semiotic potency for contradiction (i.e., feces vs. Venice), described also by Dominique Laporte (2000) in his shit vs. gold binary, Kundera notices that such facilities are evidence of hypocrisy. The more elaborate and beautiful the bathrooms, the more hypocrisy in ignoring the essential physiological processes of the body. The same logic can be applied to other types of waste, their management, and associated infrastructure. The development of waste management policies often reinforces the hypocrisy. Slick incinerators, robotic recycling centers, or experimental physical-chemical devices share a lot with Kundera's toilet bowls reminiscent of beautiful water lily blossoms designed to forget the poverty of the human body.[6] Pavel Mašek's observation that garbage trucks are ideal apparatuses, in Vilém Flusser's sense, because they deflect attention from their content and use their surface instead to build trust, resonates with Kundera's story.[7] Those who have had an opportunity to experience the interior of these trucks would probably confirm the contrast between the outer surface and the interior, including the content. The new technologies for waste management make it easier to ignore the poor garbage by emphasizing slick metal sheets, walls, nice colors, and positive slogans. The treated matter is not only out of sight, but it is believed that it can disappear entirely.

Alexander and O'Hare (2023) also make a key point that unknowing exists in a dialectic relationship to knowing. In other words, analytical attention should be paid to both these ways to "divert attention away from waste and its consequences and, in others, emphasize waste to create opportunities for dispossession and intervention" (Alexander and O'Hare 2023: 428). The need for waste's disappearance clearly goes hand in hand with its return and the need to keep it present in various senses of presence, be it its unexpected appearance in landscape, containers of waste workers, households of waste pickers, works of artists, inside bodies of organisms, in bird nests, as stench on work apparel or in car trunks, or various media representations where garbage and landfills do some performative work.

Waste is never either fully present or absent. It might be suppressed and pushed away in certain contexts, but it has a surprising capacity to cross boundaries, both imaginative and physical, and reappear always ready to engage with new relations. Even contemporary struggles to make waste disappear through circularity are problematic (Corvellec, Stowell and Johansson 2022; O'Hare and Rams 2024a). Waste's disappearance has always faced its reappearance and unexpected presence. Hetherington's (2004) conceptualization of disposal as "placement of absence" builds upon the scholarship that puts into its center of interest the relationship between order-making and rejection of certain elements (see Douglas 2002). It is not enough, however, to reject the elements through sorting, it is necessary to eradicate them (Liboiron 2019). Waste must disappear to prevent its undisciplined agency to harm, or, as Reno puts it, avoid "alternative possibilities we would rather not experience or imagine" (2016a: 10–11). This understanding of disappearance, however, is about the productive potency of absence that holds and becomes socially effective as such. But the disappearance of waste is a fantasy that is both imperfect and reversible. Waste's odor spreads, wastewater leaks, pieces fly or get transported by humans, mammals, and birds, and incineration of garbage is not without residues. Various clues and signs signal waste's victory over disappearance. It lurks around and reminds us that it cannot be gotten rid of easily.

Waste can appear in public spaces as a powerful sign that conveys various meanings. Given the fact that waste can be in more than human perspective conceptualized as a sign of life (Reno 2014), it is productive to examine what kind of lives are signified by such signs. While Reno's primary aim was to extend the conceptualization of waste beyond the human, his semiotic approach also works for the contexts when waste-as-sign gets channeled back from the doom of forgetting to shape social life. The physical appearance of garbage and even hoarding in public spaces can serve various political purposes. In the "You stink" (*tal'at ríhtak*) movement in Lebanon, garbage amassed on the streets or even sent in packages to politicians was a vehicle of resistance against the supposedly corrupted political regime (Abu-Rish 2015). Trash revolts in Dakar paralyzed the city via dumping household waste in public spaces to raise the issues of social inequality and the value of labor and served as a vehicle for negotiating collective identities (Fredericks 2018). Berg (2023) shows that garbage strikes may lead to "garbage talk," which offers actionable solutions but leaves the real causes unaddressed. Her emphasis on garbage as an opportunity to reframe debates and shift attention indicates that waste's semiotic presence is intimately tied to absence. In other words, being present may be more about hiding what is not supposed to be present.

An ethnographic immersion into the lives of landfill workers, waste pickers, employees of incinerators, managers of companies, people living close to waste infrastructures, and other organisms thriving on garbage enables me to learn and narrate what it means to live with waste in uncertain times. It offers a view into the world full of unexpected consequences, creativity, emerging relations, and forms of living to demonstrate that waste's disappearance is only illusory, as Ivan Klíma argued in the initial quotation of this chapter. It is not only an issue of the slow decay of materials that resist disappearing (Hage 2021: 3; McDougall 2021) but also waste's capacity for engagement (Gille 2007; Henig 2019; Hird 2012). As Sophia Stamatopoulou-Robbins argues, waste never truly disappears; it only changes place and form (2020: 23).

In this sense, the project builds upon existing scholarship concerning the treatment of primarily municipal solid waste through ethnographic immersion into the lives of those who care for garbage on behalf of others (Butt 2023; Doherty 2022; Fredericks 2018; Furniss 2017; Gutberlet et al. 2021; Millar 2018; Nagle 2013; O'Hare 2022a; Reno 2016a; Resnick 2021; Saethre 2020; Stamatopoulou-Robbins 2020; Zapata and Zapata Campos 2015). These ethnographies demonstrate that living with garbage is anything but passive witnessing of its disappearance. In contrast, garbage offers a multitude of opportunities for engagement, including those presented in this book.

TECHNOLOGIES FOR DISAPPEARANCE

Czech Minister of Environment Petr Hladík articulates explicitly what seems to be the most pervasive approach to landfilling in Czechia: "We should completely eliminate landfilling ... The landfill ban is insurmountable; without it and others, we will not move forward in the circular economy and in meeting the European target."[8] Although the pan-European plans are less radical, they lend landfilling a special role in the story of waste's disappearance. Landfills literally represent disappearance squared. Not only does garbage need to get out of sight and mind, but the landfills themselves should also disappear. Disposal at landfills[9] is often perceived as outdated technology that ranks at the bottom of the preferred means of waste management in the European Union (EU 2018b). By 2035, the member states should reduce the landfilling of municipal waste to 10 percent of the total amount of municipal waste generated (EU 2018a). The European Commission made an ambitious step forward to reduce landfilling to the necessary minimum. After thousands of years of waste dumping, this practice is disappearing. Although EU plans and documents aiming at greater environmental and social responsibility try to be realistic, vernacular interpretations in places like Czechia push the meaning further to imagine and call for the end of landfilling. These interpretations aim at the point that Patrick O'Hare (2022b: 63) metaphorically describes as an act of throwing landfills onto the trash heap of history.

Although landfilling has its dark side, other means of waste management are not innocent either. Incinerators are well-known for reducing volumes but increasing toxicity (Gille 2007: 25) and their contribution to climate change is a question of the timeframe considered (Ballinger et al. 2020). Identifying malfunctions of these highly complex technologies resulting in pollution is not always easy to detect (Zhang 2024: 79). Recycling may have negative consequences via pretending to solve problems while serving the needs of the unproductive "busy-ness" and diverting attention from real causes of problems (Hird 2022; MacBride 2011), allowing environmental degradation (Liboiron 2009), and generating jobs that are difficult to handle (Alexander and Reno 2012; Stowell and Warren 2018; Uhlová 2018). Moreover, the materiality of waste matters. As Forbes McDougall et al. (2008) argue, different waste management options are appropriate for different waste fractions. Therefore, "there are no overall 'best' or 'worst' options" (McDougall et al. 2008: 10). The omnipresent push for circular economy does not necessarily lead to a more sustainable future either (Corvellec et al. 2022).

If waste disposal were unambiguously the worst solution for waste management, one would expect its inferiority to be evident irrespective of space and time. However, this is not the case. From a global perspective, disposal at landfills has experienced historical ups and downs (Melosi 2020; O'Hare 2022: 59). Sanitary landfills still serve as a pragmatic solution for a world dominated by mass consumption (cf. Melosi 2020, 550; O'Hare 2022b: 64; Reno 2016a: 213–15). Formal comparisons of various management means for municipal solid waste do not show a black-and-white picture and suggest rethinking waste hierarchy without assuming that one size fits all (Vaverková 2019; Vergara and Tchobanoglous 2012). Even environmental engineers argue that a loop closure within circular economic models cannot be achieved without paying attention to residues and places for their final disposal (Cossu 2018: 86). In other words, imagining a world without places to dispose of unwanted matter is unrealistic.

There seems to be something other than the mere disadvantages of landfilling that encourages the need to get rid of it. In times of crisis, societies tend to develop various mechanisms to cope with it. As David Edwards (2017) argues, one of the solutions to a crisis is giving up something. For Edwards, giving up is a social mechanism that handles the danger coming from chaos-producing crises. Giving up entails a whole spectrum of acts ranging from abandonment to violence. René Girard (1986) envisioned crisis as a trigger for collective violence directed toward those who could be blamed for causing the crisis. Although the classic models of scapegoating work with living creatures, the logic of blame has a seductive charm that can expand to other realms that have an intimate relation to life. Using disposal at landfills as a scapegoat has the potential to make the world meaningful again and secure continuity. Although the EU experience varies from the discourses on the reduction of landfilling (Corvellec and Hultman 2012) to those calling for its ban and entire elimination, this means of waste management is imagined as a major culprit in the story of human failed attempts to wrestle with waste.

The role of crisis as a trigger for giving up deserves attention. Although Girard's (1986) original notion of crisis mobilizing the collective response was associated primarily with "social crises" related to the dissolution of social differentiation or social order in general, the spectrum of crises that may cause similar effects is wider. Today's world does not seem to be understood well via the singularity of a crisis. In contrast, multiple related and untangled dimensions of the crisis and smaller crises at different scales constitute what Thomas Eriksen (2016) calls overheating. Considering the individual dimensions of the crisis, such as environmental degradation, financial collapse, and global inequalities, separately does not augment but instead fragments our understanding of the problem (Hornborg 2016: 18). These connections, frictions, and interdependencies may be best viewed as polycrisis (Henig and Knight 2023). Capitalism is pushed to set realignments in motion when crises emerge and chaos spreads. As Anna Tsing (2005: 42) points out, such realignments may be associated with the calls to give up certain regions to enable the continuity of capitalism itself.

Waste and crisis are the usual suspects in the narratives about the world falling apart. Kelly Alexander (2024) describes the role of food waste in triggering moral panic. Building upon the ideas of Stanley Cohen (1972) and Stuart Hall et al. (1978), Alexander analyzes public messages featuring food waste in Belgium and the EU more broadly to demonstrate their potential in generating a sense of urgency. Messages concerning food waste seem designed to highlight a supposedly key issue and prompt the public to act. Alexander shows that, however, moral outrage redirects attention to consumer responsibility, while the excessive choice enabled by the market remains unaddressed.

Anne Berg (2023) addresses the close relationship between waste and crisis. Not only does waste represent a symptom of crisis, but it also serves, paradoxically, as a solution. Berg argues that large-scale systemic crises look more manageable when attention shifts to something else: questions of cleanup and moral expectations related to garbage management. More importantly, this shift to garbage talk and garbage practices that offer seemingly easy and quick solutions to the crisis "leave the causes of that crisis unaddressed" (Berg 2023: 48). Therefore, refocusing attention on giving up waste disposal in the face of technological advancement is a well-tested strategy to shift attention from societies' main problems. In the case of European waste disposal, however, garbage makes another step forward. Embodied by monstrous masses of unwanted matter that fill specific parts of the landscape, landfilling itself, as practice, should disappear. This practice is given up for the greater good and the continuation of life.

Landfilling has a crucial quality that makes it prone to be given up. In a time of increasing environmental awareness and scrutiny, landfills embody a clear link between the profitable disposal[10] of garbage and the negative consequences of this process. Key actors in waste management tend to conceptualize the problems of garbage via negative externalities: peripheral costs resulting from a specific kind of waste management.[11] In the case of landfilling, the potential negative externalities such as leakages and fires can be linked to the particular place where the trajectories of discarded things supposedly end. Although linking is not always easy and is contingent upon slow temporality (Doherty 2019; Nixon 2011), landfills' materiality and immobility make future scrutiny possible.[12] Landfills are out there with their full bellies firmly set in the landscape.

Both incineration and circular models offer better opportunities to decouple profit from its negative consequences. Incinerators are not places where matter ends. They are rather highly sophisticated black boxes for transformation where garbage enters and residues leave.[13] The residues are mobile and continue their journey to uncertain "somewheres." Circular models are even more capable of befogging the link between profit and negative externalities. Circularity may slow down the negative environmental impacts, including their very recognition (Corvellec et al. 2022: 6; Millar, McLaughlin and Börger 2019). Things move, get reclassified, have negative consequences, and responsibilities get fragmented and distributed. Leaving landfilling behind is another step in the story of minimizing obstacles to generating profit at a time when industrial activities face the danger of being subject to increasing scrutiny. As Alf Hornborg argues, "the very rationale of capitalism is to keep such externalities external" (2016: 115). The more uncertainty in linking negative externalities to profit-making, the better for the economic subject that uses waste as a means of profit. Landfilling is too straightforward to efficiently obscure socioeconomic relations and hide its material consequences.

One of Marx's major contributions to understanding capitalism was a recognition of its mystification powers. Although the concept of fetishism has appeared at the center of attention concerning mystification (e.g., Ellen 1988; Gudeman 2008; Taussig 1980), Marx's more general aim was to depict the entire system as effectively obscuring the basic principles and relations upon which it relies. As David Harvey emphasizes, a specific method enabled Marx to "discover how deceiving the world of appearances can be" (2010: 8). Marx's analysis uncovered the capitalist power to mystify social relations and their effects on the economic process. Representations of the capitalist system, both in public discourses and theorizing of pre-Marx political economists, failed to acknowledge the social constitution of capital and its inherent potential to generate social inequality. For Marx, the mystification is "implicit in the relations of capital as a whole" (1990: 1024). Although the critical idea of focusing on what can be hidden behind overt appearances of any socioeconomic system offers rich space for examining mystifications, opacities, or deflections, I will direct the reader's attention to a narrower problem of the development of this trend.

Market economies tend to develop in less and less understandable forms. As Stephen Gudeman (2016: 23) argues, one of the dominant trends in market economies is their expansion into abstract spheres. This abstraction increases the distance from the material world and has other effects. Abstraction combined with sophisticated technologies enabling high-speed trading or decentralized forms of value exchange and storage generates space for discovering new ways of profit-making. A limited ability to understand these abstractions and trace their inner workings prevents recognition of their adverse effects, such as failing to offer new social alternatives after the demise of the state (see Tremčinský 2022) or their environmental consequences.

Neoliberal versions of capitalism, as systems heavily relying on the calculative market logic, have successfully hidden its effects on the rise of social inequality (Harvey 2007: 119). With the increasing capacity of capital generation, the inner workings of capitalist formations have become less visible. While increasing abstraction and technological complexity generate the "smoke and mirror" effect, there is another way to reach the same end. Anna Tsing (2005: 27–28) argues that new frontiers of capitalism are contingent upon the capacity to confuse various boundaries. Such confusion effectively stimulates new opportunities for profit-making and accumulation of capital.

In the world of discard and waste, the confusion stems from the indeterminacy of waste (Alexander and Sanchez 2018; Gille 2013; Henig 2019; Hird 2012; Lepawsky 2018). Waste is not simply something out there. It is part of the social orders, discourses, and material flows operating in societies but emerges from specific social arrangements and ideas about social life and what might be left behind (see Furniss 2017: 305; Gille 2007: 18). Recently, Michelle Schmidt described the contingencies of the waste concept using her ethnography of the Mopan Maya. She demonstrates that " 'waste' is an ideational and material construct that turns people into consumers, nature into objects, and surplus into trash" (2022: 104). For my argument concerning the potentiality of waste to create opportunities for market, politics, and social relations, it is primarily the categorical and discursive ambiguity rather than the ontological indeterminacy based on the material liveliness of the waste matter itself (see Hird 2012). I will demonstrate that waste's indeterminacy always lurks around and offers various opportunities for those who can recognize this potential.

CZECH SOULS AND THE EUROPEAN PROJECT

This book focuses on Czechia. This move attempts to break from the grand narratives about Eastern Europe as a uniform block contrasting with Western Europe. These monolithic views of the social world in Eastern Europe create a false image of a universal post-socialist experience (Caldwell 2009: 4). Although certain features of social life and structural conditions resonate well across Central and Eastern European countries affected by Russian imperial presence before 1989, there are differences in the ways individual countries have adapted to the Europeanization project (Jehlicka 2002: 18). Capital accumulation can affect spaces differently and challenge a rigid notion of geography based on fixed centers and peripheries in Europe (Kojanić 2020).

Living with and relating to waste is a productive arena for understanding the nuances of the Europeanization project because it touches upon deeply rooted assumptions and values of what the appropriate modes of engagement with discard should be like. This arena has the potential to reveal the specificity of practices and modes of thinking in this country as well as similarities shared across larger European spaces.

One of the arguably best portrayals of Czechs is in Mariusz Szczygieł's novel *Gottland*.[14] This Polish novelist provides a nuanced and sensitive immersion into the history of Czech society using the life stories of selected people. The tragicomic book builds upon Franz Kafka's playful approach to polysemy and absurdity to depict the coexistence of seemingly incompatible social elements. An expansion of Baťas capitalist shoe empire before 1948, the greatness of artists, and the bravery of those who challenged the Communist regime are juxtaposed with the collaborators with the Nazi regime, builders of the monuments to demonstrate the never-ending friendship with the Soviets, various culture-specific silences, or unexpected consequences of the postsocialist transformation after the Velvet Revolution. Szczygieł's journey into the Czech souls is helpful in providing a sense of how contradictions and absurdities offer a framework for social life. Contradictions represent an essential feature of all societies (Berliner et al. 2016), but Czechs seem to have this logic embedded deeply under their skin. When outsiders wonder why an emphasis on formal rules and disciplining does not fit an omnipresent willingness to transgress the rules, why one of the strongest pro-Israeli societies brings up children listening to rough antisemitic jokes, why Václav Havel could become a legend while his texts remained unknown to the vast majority of the Czech population, or why Milan Kundera, the most celebrated novelist, held a ban on translations of his French-written books to his native language, or why Czechs portray themselves as overachievers in recycling while sending most of the collected plastics to incinerators (see Sosna, Stehlíková and Mašek 2024), an image of the soldier Švejk, the ultimate trickster, emerges on the scene. Should one be surprised to learn that the Czech publishers of Szczygieł's book were threatened to be sued for supposedly illegal use of the title *Gottland*? Not really. It sounds precisely like another chapter from Havel's absurd drama.[15]

The Europeanization project, which has strengthened substantially since 2004 when Czechia joined the EU along with seven other Central and Eastern European countries, brought new challenges. On the one hand, the spread of EU policies and regulations from the center has not always been smooth and their impact has not been uniform (Martínez and Beilmann 2020: 1349). On the other hand, the new countries have discovered limits in their capacities to shape the debates in the center. The project of Europeanization has had a strong disciplining component that forced the new member states to figure out ways to adapt to various regulations (see Barry 2006; Dunn 2004: 163; Gille 2016; Martínez and Beilmann 2020; Skalník 2018). These regulations, nonetheless, have been subject to creative interpretations, resistance, or misuse. Although these capacities developed during the period of state socialism and experience of life under Marxist-Leninist ideology, they continue to shape social life (see Hann, Humphrey and Verdery 2002: 11). They seem to have much deeper roots that pre-date 1948. An extensive scholarship on postsocialist Eurasia (see Balaš 2024; Berdahl and Bunzel 2010; Bridger and Pine 1998; Burawoy and Verdery 2000; Cervinkova, Buchowski and Uherek 2015; Dzenovska 2018; Hann 2002, 2016; Mandel and Humphrey 2002; Pehe 2020) tackles various aspects of the social condition. What the socialist period probably reinforced was a framework for imagining and thinking about the regulations as tools emerging in a formal system that is not necessarily compatible with practices and sentiments of everyday life. As Ladislav Holy (2003: 213) argues, the experience of state socialism resulted in an untenable tension between the images of the evil state and the great nation, which eventually resulted in the fall of the Communist regime. Although this argument captures only part—symbolic and identitarian—of the dynamic behind the fall of the Communist regime, it is essential because it suggests where the distrust of any formal structure, such as state or union, comes from in Czechia. With a certain degree of sensitivity to local nuances, a similar logic probably can be traced in other postsocialist countries in Central and Eastern Europe.

The lack of trust grows from long-term experiences of disillusion and disappointment. Every time I talked to my Romani buddy Libor, a fellow waste picker, about possibilities to improve the quality of life of his family, he listened. Still, his calm eyes told me clearly how much hope he had when a white educated *gadžo* (non-Roma) like myself brought such suggestions. He had been around long enough to believe such fairy tales. I was sad, but over time I realized that it was not difficult to scale this experience up and see a similar pattern of distrust among the Czechs related to the broader category of Westerners. The 1938 Munich agreement, when Czechoslovakia was betrayed by its allies France and Great Britain, is deeply ingrained under the skin of Czechoslovaks. A recent 2025 reenactment of a similar humiliating ritual, this time between the political representatives of the United States and Ukraine, happened in Munich again! The parties slightly changed but the inner logic remained the same: Eastern Europeans can be thrown overboard.

Can we trust again? It is hard to say, but the feeling of being EU citizens of the second category has never faded out. Opportunities to travel and test the quality of identical brands in other countries and contemplate the notion of justice behind different wages at the same multinational corporations in various EU countries do not alleviate the doubts. Moreover, centuries of Western imagination of Eastern Europe as a barbaric, uncivilized zone (see Wolff 1994) can hardly be erased during the mere two decades of shared lives within the EU. The Orientalizing mindset still permeates the thinking of politicians, laypeople, and even scholars who have supposedly high capacities for reflection (see Buchowski 2006; Jehlička 2021; Kürti 2008; Thelen 2011). A revelation that Czechia has served as a dumping ground for both legal and illegal waste (Billig 2019; Lozoviuk 2012) imported exactly from those countries that have positioned themselves as European environmental leaders and "teachers" of the barbaric East can hardly be viewed as anything else than hypocrisy.

The economy has structured the relations of Czechia to the outer world substantially. Thanks to its industrial history, Czechia developed strong economic ties, especially with Germany. In the language of managers of multinational corporations, Czechia falls among the "best-cost countries" in contrast to "high-cost countries." This euphemism is not only telling, but it also shows a discursive tool for neutralizing tacit hierarchies. Czech industrial history has been an opportunity for the heavy hitters in the West, and the state has been generous to foreign investors by supporting "investment-promotion machines" (Drahokoupil 2009: 150). This strategy not only results in negative environmental, economic, and social consequences (Delanová 2015; Franc and Nezhyba 2007; Lupták Burzová 2019; Pavlínek 2014; Schling 2017; Sosna and Brunclíková 2019), but also reinforces the notion of hierarchy and sense of being used. Waste management did not diverge from the general trend of foreign investment. Austrian, Danish, French, and German companies recognized an opportunity to grow their business and started expanding into Czech waste management from the early 1990s. The largest share of the market, however, is still controlled by the company AVE, which has a complicated ownership structure but still has a component of local Czech businessmen and women.

European concerns about sustainability and associated environmental policies and regulations were essential features in the historical development of the EU (see Barry 2001: 75). The EU has been developing agendas related to sustainable development and officially adopted them in 1992 when the Maastricht Treaty was signed. Formulating and implementing policies that would profoundly change European capacity to face systemic challenges in pursuit of prosperity while respecting environmental limits is considered one of the core values of the EU (EEA 2021: 9). In the sphere of waste management, the early policy (75/442/EEC) of the European Economic Community (EEC) specified that "the recovery of waste and the use of recovered materials should be encouraged to conserve natural resources" (EEC 1975: 2). Subsequent development led to the formulation of the waste hierarchy and the promotion of circular economic principles. Consequently, disposal at landfills has started disappearing but at a variable pace. While countries such as the Netherlands or Sweden have been leading the abandonment of disposal at landfills (Corvellec and Hultman 2012; Egüez 2021: 5), countries to the east of the Iron Curtain followed a different pathway that still relied on landfilling (Gille 2007; Weber 2022a).[16] Nonetheless, it is essential to emphasize that compliance with the waste hierarchy goes beyond the mere diversion from landfilling (Egüez 2021: 1). Also, it is worth noting that the 2018 amendment of the EU Waste Act enriched the debate about the future of landfilling via a focus on materiality. The act established specific restrictions on landfilling of recyclable or energy-recoverable materials (Zhang et al. 2022: 11). In other words, it is not only about how one may dispose of things but also about what kind of things can be disposed of. This kind of policy development resonates with Gille's (2016: 131) claim that its efficacy stems from its focus on the materiality of economic activities.

The adoption of EU regulations and policies is a story of power asymmetry. As Gille (2016) described convincingly, the unruly terrain of power frictions in Hungary-EU relations was structured by policies targeting material conditions of social life. The foundations of the EU-wide regulations were originally designed to match the needs of the producers in the countries that were early EU members. This resulted not only in disadvantageous shifts on the side of the producers in the later EU member states whose traditional ways of doing things were different—not necessarily inferior—but it also, paradoxically, enabled lowering the standards in some cases. Gille's depiction of the gradual materialization of EU politics and its consequences for the prolongation of inequalities between the former East and West fits nicely into a wider picture of "Eastern Europe" being understood as a recipient of knowledge and goods, a region in the never-ending process of improvement or catching up, a kind of blind spot on the map (see Buchowski 2004; Fishberg, Larsen and Kropp 2023; Jehlička 2021; Müller 2021). Given the continuing inequalities among the EU countries, regardless of how much they are embedded in the relations on the ground or just perceived by those who feel that they pull the shorter end of the rope, one should not be surprised that a response is often resistance, because it is precisely the last resort that the weak use to curb the strong (see Scott 1985). Moreover, going back to Szczygieł and his depiction of playful absurdity, one may realize that resistance can take advantage of this capacity via the verbal exaggeration of EU's regulations, while doing everything to bypass them in everyday life on the ground using various tactics of informality that have long tradition in the former Soviet block (see Brković 2015; Henig and Makovicky 2017; Knudsen and Frederiksen 2015; Ledeneva 1998; Makovicky 2018; Morris and Polese 2014; Polese 2023).

Ethnographic insights into discard practices and waste management in Central and Eastern Europe received arguably limited attention compared to the attention they received in Africa, Asia, or the Americas. Gille's (2007, 2012, 2016, 2025) works provided a detailed view of the Hungarian waste regimes and situated various findings in broader theorizing about the nature of socio-material entanglements, global connections, and reflections on socialism. Anthropological approaches were usually immersed more into local ways of life with waste and various salvaging activities of the urban poor (Chelcea 2015; Hejnal 2013; Rakowski 2016, 2017; Vašát 2014, 2023). Special attention was devoted to food waste, especially to the activities of dumpster divers (Kliková and Brunclíková 2017), everyday practices of food management in households (Podjed and Polajnar Horvat 2020), and thrifty activities that minimize wastefulness (Sosna, Brunclíková, and Galeta 2019). Romani studies have produced several reflections of economic activities (see Brazzabeni, Pereira de Cunha, and Fotta 2015a) associated also with sanitation, scavenging, or recycling (Dohotaru 2013; Dunajeva and Kostka 2021; Filčák and Ficeri 2021; Pulay 2023; Resnick 2018, 2021, 2025; Saethre 2020). Some studies focus on garbage and its potential to convey various meanings (Kacperczyk 2021; Sosna 2017) and provide a source of information about consumption (Antošová et al. 2024; Bilska, Tomaszewska, and Kolozyn-Krajewska 2024; Brunclíková 2017a; Jehlička, Veselá and Kubíčková 2025; Kormaňáková, Remešová, and Vančová 2021; Kubíčková, Veselá, and Kormaňáková 2021; Sosna and Brunclíková 2019; Veselá et al. 2025). Although landfilling and life at landfills represent a small portion of these studies, those that address these topics suggest that the discussions about technical and political aspects of waste regimes are reductive. They often fail to acknowledge the dimensions beyond the general idea that landfills are primarily infrastructures to make waste disappear. What life at landfills shows are diverse ways of waste's reappearance, mobilization, and reuse.

OUTLINE OF CHAPTERS

Following the Introduction, the first chapter describes the ways in which the knowledge presented in this book has come into being. It narrates the story of a researcher originally trained to analyze material remains of past societies and his unexpected encounter with the richness of social life associated with contemporary waste infrastructures. The chapter explains the origins of ethnographic research and its specificity, given a patchy back-and-forth nature in the country where the researcher was born and spent most of his life. At the center of the epistemic interest is a question of how to examine the nuances of the *familiar*. Building upon Marianne Lien's (2015) call to attend to the concepts and assumptions that are widely shared and taken for granted, I present a research strategy based on the use of the self as a vehicle for understanding the limits of the familiar. Switching between environments provided me with the opportunity to contrast and confront divergent forms of living in ways that could generate reflections on those elements that were taken for granted. Additionally, the chapter outlines the key sites, their unique characteristics, and the reasons they became central to the knowledge presented in this book. Special attention is paid to trust as one of the central components of the research of waste underflows.

The second chapter takes the reader into the world of landfill workers, who, in addition to their formal duties, interact with garbage in multiple ways. The chapter demonstrates a commitment to not letting garbage go to waste but to developing various ways to salvage it. Using Cleo Wölfle Hazard's (2022) notion of *underflows*, I examine workers' practices as an unrecognized part of the official waste streams. These invisible underflows contain not only matter but also ideas, signs, discourses, and pollution that can unexpectedly resurface to make garbage present again. Scavenging and tinkering enable workers to pull their "fish" from the waste stream, allowing them to flow into new waters. The chapter describes different regimes within these underflows that engage with garbage through reuse, sharing, gift-giving, and both collective and individual forms of commodification. The paradox of these salvaging activities is their self-evident presence, combined with their capacity to remain hidden due to being unrecognized. This creates ethical friction between workers and management, which stems from a need to conform to formal rules while tolerating salvage activities that have multiple positive effects for all the actors involved. To understand the fluid nature of this *recovery ethics*, I place at the center the notion of ordinary ethics, which accounts for the ambivalences and contradictions of everyday social practices.

Staying with landfill workers, the next chapter examines the landfill leachate's recirculation as a form of circular economy. Recirculation is a technoscientific solution to the problem of wastewater at landfills, which attempts to keep the fluid in motion within the constraints of the waste infrastructure. Landfill leachate is arguably the most critical matter at sanitary landfills because it challenges their primary function: to contain the waste. Leachate emerges in quantities that fluctuate, its toxicity is unpredictable, and its fluidity and causticity make waste infrastructures susceptible to leaks. Recirculation, practiced by landfill workers, represents the most common solution to the indeterminacy of leachate. This chapter approaches recirculation of landfill leachate from the perspective of circularity to understand the relationship between the circle as an abstract ideal and the circle as an economic model. This approach explores what forms of representation the circle engenders and what consequences circular dynamics produce. I argue that spraying the leachate back to the surface, letting it percolate, capturing it in the sumps, and eventually spraying it back again is a *vehicle of postponement*. Recirculation saves the company's money for special treatment of hazardous waste, minimizes the probability of leaks, but keeping the fluid in motion postpones the solution of what to do with the unwanted matter.

The fourth chapter describes the salvaging activities of waste pickers who shape the disappearance and reappearance of garbage at landfills in multiple ways. These people, who are dominated by the Roma, disrupt garbage disappearance inside landfill bodies and make parts of it sensible again. Building upon Kathleen Millar's (2018) *forms of living*, I examine why waste pickers do what they do despite having other work opportunities. All the challenges and obstacles associated with their state of being unwelcomed but tolerated at landfills are outweighed by freedom, the social nature of work, and an opportunity to take advantage of the waste pickers' strengths. I pay attention to the various modalities of the disappearance and reappearance of garbage. The first one is based on the disruption of the normative trajectories of garbage. Waste pickers create underflows within these streams and use tactics to temporarily hide or reveal recovered items, primarily scrap metal. These practices are immersed in the fluid relations among the actors who react to the changing composition and magnitude of waste streams. The second modality relates to waste pickers themselves. While affecting waste streams and rendering garbage visible, usable, and exchangeable, waste pickers themselves remain elusive. Their existence and contribution to sustainability and the circular economy remain unrecognized. The third modality includes material sediments of scrap metal processing left in the woods, as well as smoke from burning scrap metal components. The semiotics of fragments in the soil, the positioning of knives and rotten pieces of meat to deter potential visitors, and smoke as a prominent indicator of informal scrap metal processing reveal a rich spectrum of signs that make the practices of scavenging and processing present. The last modality depicts mixing as a powerful tactic of concealment. Using two cases, I demonstrate the role of mixing in reducing an observer's ability to distinguish between different components that constitute the whole.

The fifth chapter explores the logic of organizing entities perceived as uncanny or strange. Using the Czech vernacular word *podivnost*, I examine entities that share the capacity to irritate or disturb. Building upon scholarship about classification and the role of invisible forces of categories in shaping social life, I place the center of interest on sites of discard, military areas, industrial farms, mines, and places of tragic events. Via focusing on the histories of these places, the chapter reveals an inherent logic that I call *magnetism of strangeness*. The use of a metaphor of magnetism points to the capacity of strange entities and activities to attract each other along a spatiotemporal continuum. I argue that strangeness sticks to places and tends to perpetuate itself through a series of magnetic relations over time. The metaphor of magnetism is examined from the perspective of its capacity for both attraction and repulsion, regularity of patterning, and the invisible nature of magnetic relations, which are not seen, but their consequences can be felt. To demonstrate the power of magnetism, the text embarks on a journey to narrate the histories of three places. The first describes the transformation of a mine into a dump site, sanitary landfill, and incinerator. The second place traces the history of a former military training ground used by different armies over centuries, which was later converted into a sanitary landfill, then attracted an industrial piggery, and already has plans for a new shooting ground once the landfilling is complete. The third place witnessed a rich history of mining, its decline, and the atrocities of World War II, which were brought about by the rubble of abandoned mines, subsequent dumping, and sanitary landfilling. All these uncanny histories demonstrate the spatiotemporal and classificatory qualities of disappearance as well as a position of discard and waste within a wider field of entities and processes that provoke and disturb.

The Conclusions chapter of *Europe's Disappearing Waste* brings together the disparate ideas, people, places, and histories that the reader encountered during their journey through this book. It demonstrates the local pervasiveness of a pressing need to make garbage disappear and the failures to achieve it. This part of the book summarizes the different registers and modalities of vanishing and reappearance. Although the thrust of the book rests in local conceptualization and understanding of life vis-à-vis (un)wanted matter, the Conclusions situate this particular knowledge in a wider field of practices designed to make garbage disappear in other parts of the world. I argue that the future of our relationship with waste should not be driven by the desire to make its disappearance more efficient or absolute. I argue that the better direction would be accepting the necessity and presence of waste that requires care and offers opportunities for reflection and support of alternative forms of living.

NOTES

1. The Smoke landfill (pseudonym) is a large sanitary landfill in West Bohemia used for the disposal of municipal solid waste.
2. It is difficult to calculate the financial value of the pile exactly, given its unclear size and internal heterogeneity. However, small rolls of sandpaper (0.15 x 50 m) cost ca. €40 in the store. The most common rolls in the pile were ca. 1 m wide. Even the most conservative estimate of the financial value is breathtaking. The pile could be worth at least €2,000.
3. At that time, we had no idea that, nine years later, Czechia would experience the first real tornado in its history, which significantly changed local perceptions of climate change.

4. Here, one can see the analogy with the first law of thermodynamics asserting that the total energy of the universe remains constant. Energy does not spontaneously appear or disappear, it only flows and undergoes transformation among different kinds of energy.
5. Kundera's take on waste has already received attention in scholarly debates (Kawa 2016; Žižek 1997).
6. In this respect, Copenhagen's incinerator Amager Bakke with an artificial ski slope on its surface is probably the most illustrative example.
7. Mašek, Pavel 2023. "Rusting Away: An Ethnography of Totaled Cars, Informality, Capitalism, Waste, and Sustainability at a Bohemian Auto Salvage Yard." Ph.D. dissertation, Department of Anthropology. Plzeň: University of West Bohemia.
8. This quotation comes from an interview with Petr Hladík published in the magazine *Odpadové Fórum* 2023/5: 4–5.
9. In the official EU documents, the term disposal works as a synonym for placing waste at landfills. In social theory, disposal has wider meaning, referring to the dealing with absence in social relations. For example, Hetherington (2004) develops several subcategories of disposal that do not neccessarily have direct link to landfills. Therefore, I prefer to use the term "landfilling" or "disposal at landfills" to specify clearly what I describe.
10. There is little doubt about the economic power of landfilling. It is still the most frequent means of solid waste management in industry whose global market size indicated $1.3 trillion in 2022 (Alves 2023).
11. They miss the point raised by Liboiron and Lepawsky (2022: 23) that waste and pollution are not anomalous but inherent parts of contemporary economic systems. In this sense, negative externalities are products of power relations that shape ideas about what, when, where, and how can be externalized.
12. An example of this is a police investigation of inappropriate disposal of toxic waste (barrels with styrene resins) in the Celio landfill near Litvínov. The barrels were disposed of without prior thermic treatment of the resins. Forensic experts require to recover the barrels and investigate also the material into which the barrels were buried. Retrieved 15 October 2023 from www.seznamzpravy.cz/clanek/domaci-kauzy-vysetrovani-zlocinu-s-toxickym-odpadem-se-protahuje-sudy-zustavaji-v-zemi-237375#utm_content=ribbonnews&utm_term=styren&utm_medium=hint&utm_source=search.seznam.cz.
13. Cf. Mašek, Pavel (2023) "Rusting Away: An Ethnography of Totaled Cars, Informality, Capitalism, Waste, and Sustainability at a Bohemian Auto Salvage Yard." Ph.D. dissertation, Department of Anthropology. Plzeň: University of West Bohemia.
14. The title *Gottland* refers to the museum with the same name devoted to the phenomenal singer Karel Gott, who won the national competition for the best singer forty-two times and became almost a sacred figure whose fame reached way beyond the country borders.
15. Gille's (2016: 10) research shows that humor and enjoyment of absurdities resonate in Hungary and Romania as well.
16. In the case of the Netherlands, the ban on landfilling even preceded EU-wide directive by two years (Zhang et al. 2022: 11).

CHAPTER 1

KNOWLEDGE GROWTH

> My field work was scrappy and disconnected—perhaps more so than that of other anthropologists. ... But my case was somewhat worse. The average anthropologist has some definite interest in some one aspect of the culture he is studying, whether it be in historical reconstruction, in material culture, in economics, or in functional analysis, and he will at least collect an adequate supply of material for a book permeated by his particular point of view. But I had no such guiding interest when I was in the field; ... I did not clearly see any reason why I should enquire into one matter rather than another.
>
> —Gregory Bateson, *The Naven* 1958[1936]: 257

Gregory Bateson's honest reflection on his research is almost a century old but still represents a provocative statement about the anthropological research design and methodology. His epistemic openness and lack of rigid methodological framework made him doubt the value of his own contribution. Time, however, has proven to be the best evaluator of scholarly work. Bateson's inductive approach enabled him to produce one of the classic books that has stimulated the thinking of generations of scholars, despite the apparent lack of initial focus.

I bring up this case not to nurture anti-scientific sentiments or marvel at methodological sloppiness. I rather ask what happens when a researcher appears in the field and allows themself to be absorbed by the everyday buzz, which brings relevant research themes to the surface as the researcher floats in the stream. In the era of grant funding and audit culture, this attitude may sound like an anachronism. The perception, however, might change when a story of an ethnographer searching for space to conduct their ethnography gets reversed. It was ethnography that found me, rather than vice versa. Between 2012 and 2022, an urge to understand life at Czech landfills lured me there as well as to other sites related to waste management. Places where garbage gets processed or disposed of are so rich in terms of social processes, unexpected relations, and sensual stimuli that one inevitably becomes curious about what the world behind the fences and gates might be like. I spent a total of fourteen months at these sites, observing, discussing, and participating to learn about the everyday lives of people as well as their nonhuman companions, their relations to waste, and the uncanny histories of places where these entangled lives unfolded. In addition to that, I spent countless hours digesting various texts related to the workings of the waste industry in the region and beyond.

My research demonstrates what contemporary *anthropology at home* may contribute to anthropology. Instead of relying upon a single long period of immersion in fieldwork, my work was divided into short research periods over several years. It was patchy, with a blurred boundary between the field and home (Günel, Varma and Watanabe 2020). The patchiness, however, enabled me to cultivate patience, reflexivity, and bring the element of surprise to social relations every time I selected days that were not necessarily favorable to some of the actors. The breaks were also useful for the recognition of changes, in terms of both the contrasts between the visits and my ability to monitor long-term changes. In the following text, I describe the genealogy of the research, the key sites, and my approach to ethnography. A view into the black box of the research should give the reader an opportunity to imagine how the knowledge presented in this book emerged and to judge the weight of my arguments, given the methodology applied.

BEGINNINGS AT THE PURELAND LANDFILL

Around 2011, I was developing ideas about a research project that would enable me to take advantage of my archaeological training while using its potential in a research field that would resonate more with recent societal challenges.[1] Along with my doctoral student, Lenka Bruncliková, we started what later became the "Pilsen Garbage Project" (Bruncliková 2017b; Sosna and Bruncliková 2017; Sosna et al. 2019). The initial idea was to build upon William Rathje's (see Rathje and Murphy 2001) garbological research and enrich it with more recent theoretical perspectives and new research tools, generating knowledge about a society that differed in several respects from the one that Rathje with his colleagues researched for a few decades. One of our goals was a stronger emphasis on ethnography as an approach that would get us closer to not only how people talk about discarding objects but also how they live their everyday lives. We believed the lack of ethnography was one of the weaknesses of the traditional garbological research.[2] In 2012, we started a small project that combined the garbological (waste composition) analysis of municipal solid waste and research among its producers. Since there was very little experience with this kind of research in Czechia (Benešová et al. 2008; Kostkan et al. 2010), we had to develop a substantial part of our methodology from scratch and keep our eyes and ears open to new impulses. Conducting the garbological analysis, nonetheless, had an unexpected consequence. Sifting through hundreds of kilograms of garbage over the summer can be done only in places where one can store the material and set up a research site for sorting, weighing, describing, and recording. The landfill was a natural choice. From the first year of the project, we came into close contact with people in waste management. To obtain permits, it was necessary to negotiate with the managers of waste management companies and sanitation workers who helped us with the collection and transport of garbage. More importantly, however, we spent several weeks during the summer at the landfill, where we interacted and became friends with the landfill workers.

Already during the first research season, I started observing the activities of workers, conducting informal interviews, and making regular notes in my notebook about life at the landfill. Unsurprisingly, I became attracted to the everyday buzz at the landfill—there were so many unexpected events from the very beginning that I could not help wondering what they meant. The following vignette demonstrates this:

> It was early August 2012. Spending another day at the Pureland landfill[3] seemed more bearable than the previous day because light rain in the morning cooled down the surface and reduced dust. Even the odor was less intense and flies stopped bugging us. It was much easier to sort garbage items on the sieve. During a short break after finishing the last load and cleaning the sieve, I took a bottle of water and stepped a few feet outside our tent to see what was going in the area where the trucks deposit their garbage. An unusually large truck captured my attention. It was not the regular garbage truck but one of those that transported various kinds of cargo, not necessarily waste. When the driver tipped out the load, it looked from a distance like a large colorful pile consisting of small pieces of something. My colleague Lenka also tried to discern what was lying there but neither of us was sure. Suddenly our contemplation was interrupted by one of the workers rushing to the pile. When he reached the pile, he lifted out something small and then did something unexpected. He put it into his mouth and started chewing. Lenka and I looked at each other and neither of us was initially able to say anything. I just felt that she was also wondering whether what we had just seen was real or we had a hallucination caused by the landfill gases released from a "chimney" located a few meters from us.[4] When we finally started talking about this shocking experience, we saw the guy emptying one of the large plastic trash bags from a nearby pile of the usual garbage and start collecting the items into the bag. We decided to go closer.
>
> When we reached the worker, we saw a uniform pile of transparent plastic bags with little gummi hamburgers and eyes inside. We were staring at what we estimated as almost a ton of candies that were wrapped in plastic bags. They looked perfectly

> fine to us, but somebody had decided to dispose of them at the landfill. When I looked at the worker, he turned to me with one of the gummi eyes in his puckered mouth and started rolling his eyes. He was attempting, semi-jokingly, to scare me. I was not scared but rather shocked that he ate the candies directly at the site. I was still a fresh discard researcher and did not even know the name of this worker at that time, but this situation provoked many questions concerning what the social life of gummi candies might look like. He encouraged us to take some candies too: "Don't worry. They are fine. Get some." We politely refused and started inspecting the candy wrappers. It seemed that there was a printing error, which was the reason for their discard.
>
> The information about the candies spread like flies on the surface of fresh food waste at the landfill. During the rest of the day, workers, truck drivers, and sanitation workers were coming with the plastic bags to the pile of candies and collecting them, sometimes in large quantities that were unlikely to be consumed by these people themselves. When I noticed special deals for gummi candies in a few small convenient stores in a nearby city, I had a suspicion about their origins. Unfortunately, nobody was willing to tell me more to evaluate my suspicion.

It is probably not difficult to understand why similar events and activities captured my attention. They were opening a new world of human creativity that was hidden behind the fences while also being connected to the outside. Although there was already literature concerning scavenging from diverse parts of the world (Fahmi and Sutton 2010; Millar 2008; Reno 2009; Zapata Campos and Zapata 2012)[5] and there were rumors and mentions in local novels that the employees in waste management took advantage of discarded objects, there was little knowledge about these activities in the region and how they were related to the bigger picture of waste management in Europe and beyond. It made perfect sense to learn more about everyday life at landfills and potentially contribute to the understanding of discard practices. I already had connections in the waste industry that enabled me to extend my research activities in this direction. Without the previous garbological research, which was perceived as unproblematic by actors in waste management, it would be very difficult to get into the landfills and conduct ethnographic research there.

In 2014, the research among landfill workers became central to me. Although we carried out the garbological research until 2016, whenever there was a chance and a day off from teaching, I went to the Pureland landfill to spend a day with the landfill workers. These visits were possible because all the employees knew me well, and the landfill was located not far from my home. I arrived, changed my clothes, and simply started chatting with the employees, and the manager at the gate if he was present, and then I moved to the area where the workers took care of the incoming garbage. I never pushed them to feel any obligation to take care of me. When workers were busy and my presence would be an obstacle, I just waited and observed the buzz. But they knew that they could count on my help even in relation to their official duties. Most of the time, however, workers engaged in scavenging, handling, and processing their finds, as well as what Joshua Reno (2016a: 124) calls *tinkering*: the creative acts of reuse. This was because the flow of incoming garbage was weakening over the years, and the workers had a lot of *free time* at work. All these activities benefited from another pair of hands and eyes, so I could be useful. Search and subsequent transport of potentially valuable objects required walking, lifting, and pulling. Also, I could hold things during repair, bring tools, and simply help with various little things, performing the subordinate role. This reversal of hierarchy resonated well with Carole Crumley's (1995) notion of heterarchy: the shifting notion of hierarchy changing according to the context. Within two days, I could switch between being a member of the advisory board for the president of a university to a person being ridiculed for holding inappropriately a hose while spraying leachate at the landfill. During work and breaks at the landfill we had many opportunities to chat. Although there were differences, most of the workers liked the opportunity to engage in conversations that Bruce Grindal (2011: 89) describes as *shooting the shit* for its refreshing honesty and opportunity to move beyond the politically correct tone.

When I first asked the landfill manager whether I could come beyond the framework of our summer garbological research to hang out with workers, he did not understand my intention. I explained to him what ethnographic research was about and that my interests went beyond the garbage itself. He was not excited but agreed. It was clear, however, that he did not understand entirely why his landfill would be attractive as an object of research. It did not fit the stereotypical image of anthropologists doing research somewhere in the Amazon. Given our existing relationship, he knew, nonetheless, that my intention was not to collect evidence for police or officers from the Ministry of Environment. I was only required to inform him about my visits. He informed the company's production manager, his superior at the headquarters, and I was able to start visiting the landfill to study its everyday inner workings. I conducted research there until 2017, when the landfill was shut down. Beyond 2017, I kept revisiting the site approximately every half a year to see its afterlife: the infilling of its capacity with slag from the nearby incinerator and the restoration of the landfill's surface.

NEW SITES

Around 2014, when my interests were focused on practices at the Pureland landfill in West Bohemia, I started exploring other similar sites in the region and beyond. I was curious about their location in the landscape and their histories. Also, I wanted to get a sense of variability and compare my experience at the Pureland landfill to other sites. I did not expect to experience the same degree of immersion in their everyday life as I experienced at my original site, but this part of the research proved to be helpful when I searched for new research opportunities. Until the end of 2022, I visited sixteen landfills, four recycling centers, one facility for processing hazardous waste, and two incinerators. My experience with these sites varied from a few visits and attempts to engage in any meaningful discussion with the employees to extensive experiences and close ties with the people at the sites where I spent a substantial amount of time after the shutdown of the Pureland landfill in 2017.

The second key site described in this book, the Smoke landfill, enabled me to study the activities of waste pickers. My initial experience dates back to March 2015, when the site surprised me with its unusual spatial disposition. Unlike most of the landfills I had visited before, the mound was not round or square but elongated. Later, when I examined the LIDAR data, the landfill almost looked like a giant but short snake lying in a valley.[6] When I walked around inspecting the surroundings, I realized that the landfill's tipping zone at the back of the site was well visible and close to the fences. Most landfills I am familiar with frequently use natural barriers such as dense bushes, forests, swamps, terrain protrusions, or simply distance from roads and trails to keep out unwanted observers. Here, one could get easily close to the tipping zone situated only 300 meters from the road in a sparse forest. In addition to such an unusually welcoming access, I was surprised by the large number of various piles of primarily e-waste, bike components, and plastic waste scattered outside the landfill in the woods. It was clear that something was happening here, and I wondered whether homeless people or somebody else might have been responsible for the littering. I visited the landfill a few more times during the subsequent years but was never lucky to see any activity in the woods. Since I was saturated with the research at the Pureland landfill, I left this topic on the list of potential research interests for the future.

In November 2020, the situation changed. I arrived around noon, after having spent the morning observing the transport of paper from one of the closed landfills functioning as a reloading station. When I arrived at the Smoke landfill, I found fresh piles of e-waste in the woods, and there was a bike leaning against a spruce trunk. There were three guys walking in the active zone collecting some material into plastic bags. I moved closer to the fence but remained hidden in the bushes. I could recognize that one of the guys was holding an electric cord, but he seemed to keep his hand behind his back to hide it. Also, he squatted down into the garbage mess a few times when the driver of the front-loader on the top of the hill started moving in the direction where the three guys were scavenging. I realized that these guys were unlikely to be employees engaged in scavenging, which was the praxis I knew from the Pureland landfill. They were dark-skinned, and their behavior was less self-confident. Later, they started complaining about somebody who kept stealing their finds when they left their bags in the woods unattended. They seemed angry and started moving toward me. Given their anger against somebody who was stealing their finds in the woods and myself being exactly next to their processing areas, I assessed that it was not the best time to introduce myself, and I silently vanished through the bushes back to the road and left the area. I was nervous but excited. I decided to come the next day and try to meet the guys.

When I arrived the next day, I could not see any waste picker initially, and the landfill seemed calm. I decided to explore the areas with garbage where the waste pickers probably processed their finds. After an hour of walking around, taking notes and pictures, I discovered a hidden *hustník*: a dense patch of forest that is hard to pass through. There was, however, a path leading inside. When I passed through the barrier of young spruces, I discovered a bizarre scene. There was a huge hearth full of black ash containing small pieces of unburned plastic, aluminum foil, and fragments of copper. Around the hearth, there were plastic bags, used gloves, plastic peels from electric cords, and iron or porcelain parts of various electrical devices scattered around. Two dirty knives were stabbed into the trees in a way that one could hardly overlook. They seemed to not only index the peeling of the electric cords, but also to symbolize a warning. After careful examination of the burning site, I continued further into the woods to see the nearby creek, where I found also signs of burned material and footprints on the bank. On the way back, I noticed a figure carrying a big bag and disappearing quickly inside the hidden burning site. When I saw glimmers of fire and ascending smoke, I inhaled deeply a few times to suppress the stress and slowly started moving toward the fire. A young man squatted next to the fire watching the burning electrical cords and poking into them with a wooden rod. Immediately after noticing me, he stood up to get ready for whatever might come. He explained that he had cleaned up the mess in the woods that somebody else had made and now was burning the collected material. I greeted him and assured him that I was neither a policeman nor a forester, so he could calm down. I explained that I was a researcher interested in landfills and the activities of people who salvage materials from there. The man was clearly perplexed as to what this experience was about, and introduced himself as Honza, which, I realized later, was not his first name but a fake name used to hide his identity. He let me stay there during the entire process of burning, coiling, and packing the burned copper. After the burning was finished, he rushed to get away quickly. He did not even wait for the copper to cool down, but took the bundle on a wooden stick instead and walked away.

That day I not only met the first waste picker from the Smoke landfill and saw the burning process, but I also met his friend Fero, who became my key interlocutor during the first months of my interactions with this group of waste pickers. He sat waiting for Honza, surrounded by four large plastic bags and a basket of mushrooms. This relaxed man, who could have been my age, was more open and started sharing his stories about waste picking from the very beginning. Apparently, it was not his first experience with researchers, because he knew people from local non-governmental organizations (NGOs) who were helping families living in social exclusion. He was familiar with the Pureland landfill and some of the employees who used to work there. My knowledge of those people and the landfill enabled me to break the initial distrust and avoid being expelled at the beginning. I persistently kept visiting, until the pickers got used to my presence and realized that it did not endanger their scavenging. In contrast, they soon realized that they could take advantage of my willingness to help them during scavenging. During the next two years, I participated in the activities of these waste pickers. At first, I only waited in their workshops in the woods, guarding the finds, helping with disassembling, and chatting with waste pickers during the breaks in the collection. Over time, I was allowed to participate in scavenging in the tipping zone inside the landfill. I was introduced to the two landfill workers who alternated as drivers of the front-loader, and they learned to tolerate my presence.

DOING ETHNOGRAPHY

My pathway to ethnography is probably unusual, but I am certainly not the first archaeologist attracted to ethnography. Richard Wilk's case has been always an inspiration for me. In his dissertation, Wilk vividly explains his original fascination with the past, especially the Maya civilization.[7] In May 1976, when the spring rains came early, he could not continue digging test excavations to collect a pottery sample from a newly discovered Mayan ceremonial site. Instead, he realized the potential of interaction with the living people:

> I spent much of my time sitting in a palm-thatched shelter watching my pits fill with water and talking with Lorenzo Shol, the watchman of the site. I learned more about the nearby Kekchi village of Indian Creek, where Lorenzo lived, than I did about the prehistory of the site.[8]

His subsequent intellectual venture via ethnoarchaeology led him to cultural ecology and economic anthropology, where Wilk could take advantage of his four-field training and find a middle ground between the Scylla of materialism and the Charibdis of idealism.[9] Although my trajectory was different and certainly less exotic in terms of my field research, it also started as a recognition that social life at our research site was worth exploring. Moreover, I came from the similar four-field background that taught me methodological tolerance and openness to other anthropologies. It is probably not a coincidence that economic and ecological anthropologies, which I taught for many years at University of West Bohemia, became my natural spheres of interest and enabled me to break the division between the past and present.

My understanding of ethnography grew slowly from reading ethnographies and from conducting field research myself. During my early career, when I directed my attention primarily to the past, ethnographic monographs served primarily as a source of imagination for understanding past societies. When I started conducting participant observation, I was getting a sense of the epistemological and methodological nuances of what *being there* meant and how I reflected "the process by which we know" (Hastrup 2004: 4). In the context of Central European overlaps between anthropology, ethnology, qualitative sociology, and social geography, I had to reflect on the specificity of my own approach to research and the dimensions that were most important for me, epistemologically speaking. Maurice Bloch's (1998) emphasis on the non-linguistic ways of knowing was crucial. My interlocutors talked, of course, and their utterances were useful for understanding what their experience of living with waste could be like. Beyond this level of knowing through their words, however, there was something else prominently affecting my understanding. It took time to recognize it, but certain things were not verbalized at all. For example, it went without saying that scavenging was tolerated and practiced during regular work. Similarly, nobody felt any need to explain why waste pickers came to the landfill and burned the electrical cords immediately the day after being caught by the police and receiving a fine for making fires in the woods. Also, I learned to recognize the significance of the themes that were absent in the speech. It took time and attention paid to small divergences from the recurrent everyday pattern of work and speech to recognize Michael Herzfeld's observation that "absence sometimes speaks louder than presences" (2001: 22). The absences, which I encountered, fell primarily into what Haldis Haukanes and Frances Pine (2021: 103) call *quiet silences*, referring to things that can be ambiguous, shameful, or even illegal. Some themes, such as leaks of landfill leachate from the underground sump, were not only absent but also actively restrained against resurfacing.

Spending time at landfills with people who exercise manual labor in such a rough environment, regardless of weather conditions and pollution, inevitably leads to epistemological sensitivity to the body. Bodies of workers are protected with what Reno (2016a: 40) calls *second skins*, and purification rituals are practiced to maintain the border between the domains of work and home, although the rituals and the notion of the work vs. home barrier vary. There are differences between employees who have access to infrastructures that enable them to take a shower and store their clothing at work, and waste pickers who have to clean themselves and store everything at home. Body is also a nexus of various environmental, polluting, and labor factors that affect it in various ways. Some of the factors are recognized and sometimes even feared; other factors remain mostly submerged, even while affecting the bodies of people who directly handle garbage on an everyday basis.

Scavenging itself is an embodied practice of walking around with eyes constantly scanning the surface of the ground for potential finds. It is reminiscent of mushrooming, a practice that is highly developed in this part of the world, where woods are full of people walking around with their heads turned down during the warm days after rain. Landfill workers and waste pickers pushed this embodied practice to a new level because of the intensity and length of their practice over years or even decades.

In the noisy environment of landfills, bodily postures and gestures are also important vehicles of communication. Self-confident postures serve to cover the informal nature of scavenging or threaten a rival when tensions emerge. Although it might sound like an ill-suited intellectual shorthand to project animal nature onto humans, something Marshall Sahlins (1976b) warned us about a long time ago, I could not help recalling what I had learned about locomotion and communication during my undergraduate training in primatology. I fully embrace Millar's (2020: 5) critique of the dehumanizing effect of the superficial depictions of waste pickers as animals. My perspective, however, attempts to do the opposite: to recognize the shared capacities of different species as a way of challenging human exceptionalism and acknowledging that we inhabit the world together. I am convinced that, as anthropologists, we have a commitment to acknowledge not only the social dimension but also, among other things, the biophysical dimension (Escobar 1999) that inevitably provides the basis for life and shapes it in a multitude of ways (Bateson 2000; Deacon 2012; Ingold 2000).

The postures of some workers emphasizing grandeur and corporeality served as a clear threat to the skinny manager. The same bodily tactic was used by the tallest and stoutest waste picker to protect their rights to scavenge and avert expulsions from the landfill. It was a kind of signal that simply works in a world where physical strength and body size remain important for fulfilling everyday tasks at work. The same holds for the hiding tactics. Because of the volatile relationship between waste pickers and official workers, we had to crouch and freeze in the garbage every time there was a risk of being seen by some unfriendly drivers during scavenging. This embodied practice was based on a quick crouching into the irregular terrain and waiting until the danger disappeared. One simply tried to fit into or become part of the environment full of colorful heterogenous material. It is a kind of camouflage tactic reminiscent of, for example, a rabbit crouching, freezing, and almost trying to become embedded in the ground to avoid the attention of its predators. While ethnographic literature on hunting emphasizes the intentional and well-planned camouflage used by humans to deceive hunted animals (see Bubandt and Willerslev 2015), the practice of waste pickers is rather an immediate habitual reaction to danger suited to the environment where the pickers move.

My commitment to experiencing the work of people at landfills first-hand led me to participate in the activities that they conducted. It was the only way to capture the dimensions of their work that are difficult to verbalize or that simply unfold without being noticed or reflected upon. Millar's (2018) ethnography vividly describing the experience of being immersed into extreme pollution resonated with what I felt and tried to understand in the field. Walking through the soft matter, the practice that any human scavenger necessarily undertakes, is not just about learning to adapt to bad odor, the presence of decaying organic matter, sharp objects, omnipresent flies, or occasional rodents; it also opens new registers of experience. Soft matter *vibrates* under one's feet when trucks or front-loaders pass nearby in a way that is reminiscent of listening to an incoming train with one's ear laid on a rail. The difference is that one feels it through the soles of one's boots. The softness and irregularity of terrain is physically demanding. Mountaineers know the feeling of "two steps forward, one step back" during ascents through steep snow fields. At landfills one tends to walk up and down the slopes, and a few trips while carrying a bag of finds or rolling a heavy stove makes it physically demanding. No wonder the scavengers needed breaks. Walking on the surface of garbage also requires paying attention to every step to prevent a sprained ankle, since a foot or lower part of the body can easily sink into the matter. Indeed, one has to pay attention to unstable piles on the edges of steeper slopes, sometimes reminiscent of snow overhangs, that can fall and bury a person. While I never experienced directly anybody being buried as Millar (2018: 2) described, waste pickers including myself were careful when approaching such unstable piles. The guys commented on the need to be careful, so there was a knowledge that unstable piles could be dangerous.

One also learns what it means to get the garbage literally under one's skin (Millar 2018: 56). Work without gloves is more common than with them, especially among waste pickers who are not willing to spend extra money on an unnecessary piece of equipment that gets torn and limits one's feel for the material. Interestingly, when waste pickers use gloves, it is primarily as protection from cold rather than from pollution. Most pickers I worked with considered it natural to have bare hands able to feel the material, and my questions about gloves sounded surprising to them. For me, working without gloves was an opportunity to add another dimension to my understanding of what closeness to waste might be like. Beyond the level of directly feeling various modalities of dirt—namely sliminess, stickiness, coarseness, or dustiness—and getting used to its omnipresence on the skin of the hands, there was also an important dimension of skill in tearing, lifting, or pulling things. Accomplishing working tasks without being cut or hurt was challenging and required training. In addition, working without gloves was a way to fit in and not stick out as a person with weird habits.

One of the most pressing issues for the landfill workers I worked with was liquid rather than solid waste. Landfill leachate, which gets formed in large quantities in the body of the landfill, has to be managed to prevent its leaks. The favorite strategy used by managers to deal with leachate was to recirculate it. At the Pureland landfill, this meant everyday spraying of leachate on the landfill's surface using hoses. The workers had to hold the hoses in their hands during spraying and were exposed to the air that was humidified with droplets of leachate. Spraying was particularly unpleasant during windy days when one was easily exposed to leachate droplets accidentally hitting one's face. Pollution penetrated even deeper than under one's skin; it was inhaled, and occasionally swallowed when droplets got into the mouth. Experiencing this kind of exposure to pollution was critical for understanding why the workers complained so much about spraying leachate and believed that it was the key agent behind their health problems.[10]

Skill was another important theme that I paid attention to and tried to understand via observation and practice. Although I live in a house and followed a dominant Czech attitude of building and repairing whatever one is capable of, scavenging opened new dimensions of practicing and understanding skills. For the people who spent most of their lives doing manual work, I was perceived as an impractical and unskilled academic. I experienced failures and did not always have enough imagination to solve tasks that required quick adaptation to unexpected events. I was ridiculed for my lack of skill both among waste workers and waste pickers. Although my competencies improved over time, I was perpetually behind my peers. The skills of waste pickers, in particular, were truly exceptional. Although they do not bring with them special tools, with the exception of wire cutters, they can always disassemble, divide, modify, and transport their finds. They are *bricoleurs* (Lévi-Strauss 1966) in the purest sense of the word: never deterred from seemingly impossible challenges and always ready to experiment and use their skills to accomplish a task. Without seeing it, I would never have believed that it is possible to disassemble water taps, copper pipes, stoves, or power transformers with a metal rod and wire cutters. One realizes that, instead of using a complex set of special tools and following the prescribed ways of interacting with the matter, one needs to learn where and how to hit a screw with an iron rod, lever iron parts covering the copper bundles in power transformers, or cut copper cables to be able to peel them by hand. Conducting these activities enables a researcher to learn about the necessary skills and strength, the pain, the polluting agency of the materials via touch, and the emotions associated with these activities. As Andrew Sanchez (2020) argues, it is not only the construction and building of things that generates emotion; the application of skills during disassembly generates emotions such as satisfaction as well.

Monitoring garbage that was spread beyond landfills by both human and nonhuman agents, such as birds and wild boars, exposed me to various material things distributed in space. I was able to build upon my archaeological sensitivity to things and my skills in documenting them (Olsen 2010; Olsen et al. 2012). I recorded the variability of things, their spatial position, and their use-wear, which was particularly useful for understanding the contribution of different creatures in the spread of pollution beyond the landfills. Interest in what Alfredo González-Ruibal (2008) calls *thingliness* inspired me to appreciate the fragmentary nature of things as well as the effects of their spatial accumulation or spread. Interest in materiality led me to recognize the qualities of the materials, such as flexibility, shape, color, or ability to last. At the same time, the semiotic potential of material things in carrying iconic and indexical logic of reference opened up avenues for thinking about the past and possible futures (Keane 2003: 417).

BEING CLOSE TO HOME

Doing "anthropology at home" (Jackson 1987a) has several specificities when compared to research in which substantial spatial, temporal, and linguistic kinds of "displacement" construct the field. Ghassan Hage argued that, "if one looks hard enough, otherness lies hidden amid the institutionalized sameness of one's homely spaces" (2023: 2). This kind of close otherness may even constitute the strongest forms of difference available (Marcus 1986: 168). Judith Okely aptly noted that "some of the inhabitants of one's own village may remain as elusive as cannibals" (1996: 4). The main driver causing me to approach closely located spaces of social life as my field sites was curiosity about the unknown social worlds that dwelled behind the gates of landfills, and their relation to wider social, economic, and political processes. The paradox of the unknown being so close and, at the same time, essential for economy and social reproduction was the initial trigger of my research interests. It was not otherness, however, that helped me to conceptualize what I was studying. The reason for distancing myself from otherness was not the inflationary effect of this concept for social theory (see Fabian 2006: 148). It was the recognition that people in the waste industry and I participated in the same story, despite differences in our individual lives. Instead of facing otherness, I felt a sense of being within, being part of the wider relations that connect waste streams with diverse forms of life. Garbage produced by our own household was transported into places where my interlocutors worked. We all shared space and contributed to regional history with our negligible lives. My research was permeated with a sense of mutual dependence despite its invisibility, and of shared experience despite differences. This sense went beyond the frame of the research. Research projects come and go, but people tend to stick with places where they live, especially in countries like Czechia, where this attachment is strong. The research was a way to discover, learn about, and appreciate people and processes that were almost invisible and unacknowledged but were essential for life, *sensu lato*, in the region. I believe that it fits to Catherine Verdery's suggestion to "give voice to the 'natives' as analysts of their own condition" as one of the possible ways to develop post-socialist-cum-post-colonial studies in Euroasia (Hann et al. 2002: 20).

Marianne Lien (2015) made a substantial theoretical intervention to understanding and practicing anthropology at home. She argued that difference and otherness—the champions of anthropological imagination—are not necessarily the most productive concepts to stick with when anthropologists examine the nuances of the familiar. Lien asked: "But how do we question that which is taken for granted? How do we unpack, or problematize, those entities and distinctions that seem self-evidently true?" (Lien 2015: 21). Her answer was that practicing anthropology at home requires paying special attention to the concepts and assumptions that are widely shared and taken for granted. This approach led me to go beyond the apparent interest in elements that made landfill workers, waste pickers, drivers, managers, or me different. Thinking about our being in the same world and experiencing its materiality, shared reactions, and feelings, understanding signs, and overlaps in our values directed me to search where the obvious and unreflected came from. It opened uncomfortable questions about my own biases and unquestioned assumptions.

Contrary to viewing anthropology at home as potentially less prestigious in global hierarchies of knowledge production (see discussion in Brković and Hodges 2015), it has been strong in waste and discard studies (e.g., Nagle 2013; Reno 2016a; Reno 2020). This scholarship has contributed substantially to both knowledge of specific waste regimes and the advancement of theories concerning waste and discard. It proves the point raised by Michel-Rolph Trouillot (2003) that the key problem is an epistemological distance rather than a physical one. I recognize the multiple strengths of research at home. Knowledge of language, common values, or a shared experience of historical events represent the obvious advantages of a native researcher. In my case, my family background and experience of living in an industrial environment were advantageous in tuning myself into the same frequency, especially with landfill workers. I was brought up during the 1970s and 1980s in a working-class environment in North Moravia, one of the most industrialized parts of Czechoslovakia, where I was literally immersed in the world of workers until the end of my high school years. Despite both temporal and spatial distance from such experience, I could mobilize it to relate to people and work with them, and I could juxtapose my recent ethnographic experience with what I had already known to better understand life in the waste industry.

In addition, living almost two decades in the region where one conducts research means that one knows the long-term processes not only through reading or listening to narratives of people but via one's own experience. This long-term everyday exposure to general processes in the region enables one to intensify what could be seen as a core of ethnography: "using the self—as much of it as possible—as the instrument of knowing" (Ortner 1995: 173). There are also less-expected effects of living in the region where one conducts research. I taught anthropology at University of West Bohemia, and some of my anthropology students found jobs in local NGOs. An exploration of the broader social ties of my fellow waste pickers led me to discover, to my surprise, that my former students interacted with them as social workers. The circle closed. More importantly, the students mediated knowledge that would be hard to access for any outsider.

Lack of prolonged separation—correlating with immersion—is usually understood as limiting for ethnographic knowledge production. An ethnographer is expected to break from their usual involvement in life and immerse themself in the field. But as Vered Amit argues, "the construction of an ethnographic field involves efforts to accommodate and interweave sets of relationships and engagements developed in one context with those arising in another" (2000, 6). Doing anthropology at home means that one balances various involvements that dissolve the imaginary boundary between home and the field. For me, it meant projecting my experience and involvement of being a family member and teacher onto my interactions at landfills as well as bringing my experience from landfills and confronting my family and students with it. My ethnographic research, which might fall into the category of *patchwork ethnography* (Günel et al. 2020; Günel and Watanabe 2024) or *back and forth* (Brković and Hodges 2015), was not based on radical breaks from my normal involvement in life. On the contrary, my normal life and field research intermingled. My repeated short-term visits represented a way to wrestle with the deontological ethics of the discipline (see Castaneda 2006), the requirements of employers, and especially my role as a father of two children and husband responsible for his family.[11]

While the patchy approach could take me away from the depth of immersion that separation can entail, it encouraged me to experience directly the lives of my interlocutors as much as possible. It included participation in manual labor, direct experience with various polluting agents, injuries, and weather conditions that made any outdoor work at landfills challenging. It enabled me to experience also the affective dimension, including joy, fear, or appreciation for landfill aesthetics. Although all this was insignificant in comparison to the experience of those who have been handling waste and facing its challenges for years or even decades, it gave me at least a sense of the obstacles and the kinds of experiences that life with waste entails.

Patchy fieldwork enabled me to communicate my fresh experience from the field with others and cultivate my capacity to render the unknown familiar. Unexpected questions or comments from my daughters and students forced me to rethink my experience and learn to communicate to describe the world of dumps in ways that would be digestible and imaginable for others. At the same time, the questions of my colleagues at landfills about my family and work forced me to rethink how to familiarize them with the social environment I was returning to every time I left them. Switching back and forth between the worlds was an opportunity to think critically about them and expose them to the perspectives coming from the outside. Like in Kurt Gödel's (1992) incompleteness theorem, not everything can be proven within the constraints of a single domain or system, as Gödel would probably say. One needs to step out to prove the axioms. Therefore, stepping out and bringing information from the outside is a critical moment. Unlike mathematicians and logicians, I was not motivated to prove a theory. But I shared the conviction that to know the domain better, one needs to use information from the outside of it.

Most of my research sites were located up to 70 km from my home. Such closeness enabled me to conduct daily trips to the sites and back, similar to trips that my interlocutors undertook from their homes on a daily basis. The most challenging aspect of logistics was the transport of working apparel and boots. The necessary gear is voluminous, especially in winter. In addition, it smells after use. It is challenging to take everything on public transportation, because the odor attracts attention.[12] I experimented with various means to get to the research sites, but driving a car became the primary means to get myself and the gear to the sites. Although it is not the most sustainable way to conduct research, the car was a way to fit in, and it had a few more advantages. Employees, visitors, and even most waste pickers arrived in cars that allowed some of them to transport their scavenged finds away from the landfills. I could fill the trunk with various pieces of dirty and stinky gear, as well as transport scrap metal and people whenever there was a need for it. Also, the car served as a storage space for valuable items, IDs, and even a backpack that I could not take to the field when I worked with waste pickers. A bottle of water and cutting pliers were the only things that waste pickers usually took into the landfill. Fitting in meant doing the same. To learn about the bike trips of two of my fellow waste pickers, I used a bike for a few months as well.

Advantageous spatial proximity to the field sites is counterbalanced by several less obvious obstacles. Although research at home does not require an effort comparable to Leopold Pospíšil's (2021: 21) initial negotiation with the Debei people in the highlands of Papua New Guinea, it would be a mistake to assume that entering the field is easy (cf. Jackson 1987b: 14). One of the challenges of being a native researcher lies in the absence of an aura of a foreigner. It is easy to be rejected by potential interlocutors. Humans tend to be curious creatures, and foreigners often carry an aura of the unknown that can be discovered, or sometimes even used, especially when the newcomers come from worlds that seem superior to those inhabited by the people who are studied. This bidirectional curiosity in research, when the researcher studies others while they simultaneously study the researcher, is more asymmetrical in anthropology at home. In the era of populism, researchers in social sciences and humanities are even portrayed as losers, and their reception outside the academic environment is one of being ignored or worse.[13]

Native researchers, at least initially, can be understood as representing a higher potential to reveal uncomfortable parts of everyday lives. Foreigners facing linguistic and cultural barriers may be perceived as representing less danger because of the assumed limitations. Moreover, native researchers do not necessarily disappear entirely after the end of the research and their presence may continue, causing troubles. An estimation of the researcher's capacity to reveal the less obvious is critical for acceptance in research fields permeated with a high degree of *informality* (Hart 1973) and *il/legality* (Nordstrom 2007). I was rejected many times despite having previous connections with people in the waste industry. Reno's (2015: 561) observation that the waste industry works best when its inner processes remain hidden reveals why it makes sense to keep any sniffing researcher out of the game. The cocktail of these factors does not make it entirely easy to conduct research at home, despite its seeming comfort and easy access to the field.

TRUST AND ETHICS

The names of people, places, and companies are anonymized in this book unless stated otherwise. This approach enables me to keep promises given to my interlocutors as part of the consequential ethics of my research (see Castaneda 2006). The world of professional waste management is not easily accessible, and many of its participants actively discourage anybody from the outside from getting in. There are multiple reasons for this. The waste management I am familiar with is full of informality, illegality is common, and problematic issues are often silenced. In addition, having an intimate work relation to the polluting agency of garbage may carry a stigma that not everybody wants to reveal. Being part of that world was a commitment to minimize the probability that my work would cause harm to those who were willing to tolerate me or became close to me.

It would be false, however, to describe my relationship with the interlocutors as harmonic and gradually improving over time. Although I did not conduct research in an atmosphere of omnipresent distrust or violence, there were rare occasions when I was deceived, robbed, or threatened, both verbally and physically. Some of my interlocutors spent years in prison for battery or were training to become mixed martial arts fighters. Therefore, they had a temper I had to take into account. *Fluctuation* would be the best concept to capture the dynamics of our relationships. Since my field research was not continuous, some of the relations cooled down or became warmer as a reaction to my absence or presence. It cannot be stated, however, that the *social warmth* was simply a function of continuity or length of interaction with those I studied. While the mutual trust between workers or pickers and myself generally improved with longer blocks of my research, my relationships with managers were different because of their changing anxiety about the potential consequences of the research. This was especially the case with the manager at the Pureland landfill. The more frequently I visited the landfill, the more anxious the manager became. There were times when he prohibited me from visiting, arguing that I was there "too often." It was especially during the construction of a new incinerator next to the landfill when his company became highly sensitive to the activities of activists and journalists who wanted to know more about the project. The same anxiety appeared after heavy rains, when the manager wanted to prevent any information leak about his troubles with the management of landfill leachate. I respected his decisions and waited for an opportunity to return. Patience worked well, and the manager always cooled down and granted me permission to visit again. Naturally, I am aware that some sensitive issues could remain hidden from me. Anyway, the patchiness of my research was not shaped only by my commitments but also by the fluctuating anxieties of the managers who denied me access to their landfills when it seemed too dangerous to have me around.

One of the challenges of the research was balancing multiple loyalties. Managers, workers, and waste pickers are not equals, and there are tensions among them. There are tensions within the groups of workers or waste pickers. As Hannah Fair et al. (2023: 8) suggest, the researchers inevitably ask: "on whose behalf do we speak?" Balancing multiple loyalties means that one also makes bad decisions and fails. As Carolyne Nordstrom says: "No text, no academic guide, provides straightforward maps for navigating these whirlpools and eddies of professionalism" (2007: xviii). When I was unsure about whom I should protect and on whose behalf I should speak, I was mostly inclined to protect or support those who seemed to be the least powerful. This, of course, is never entirely clear, because hierarchies are not set in stone. Still, I argue that such an approach always requires some degree of concealment, which Sherry Ortner (1995) describes as *ethnographic refusal.* Contrary to the fact that a strong sense of transparency is considered one of the key values in North Atlantic societies, I join Jerry Muller (2019: 160) in arguing that some degree of opacity and concealment is an essential part of human life.[14] This, of course, does not mean resignation from opening problematic issues and cultivating discussions concerning research ethics that bring together various actors and their opinions, the approach that Fair et al. (2023) advocate.

My relationships with landfill workers were closer than my relationships with the managers I described above. Workers were willing to share with me information about their scavenging and smuggling of scrap metal. But there were still themes that they were not willing to share with me. Stealing gas from the machines was something that I witnessed by accident, but nobody was willing to admit that it really happened. When the so-called migration crisis hit Europe in 2015 and my opinions differed markedly from those of most of the workers, our relationship was affected. Despite trying to be neutral, I could not help making occasional comments against the most xenophobic and racist arguments that were flourishing among several colleagues in waste management at that time. In addition, one of the workers, whom I call Karel, never liked the idea of my research and always kept a distance. I am afraid that I never earned his trust.

The question of trust was even more critical for my relationship with waste pickers. They were outsiders coming to the Smoke landfill through the holes in the fence, and they depended completely on the mood of landfill workers who either tolerated them or tried to evict them from the landfill, especially after the occurrence of landfill fires, presumably caused by the waste pickers, or thefts of material considered by waste workers to be their property, or before visits by inspectors from the Ministry of Environment. Also, waste pickers had to be careful when burning copper wires and cables in the woods. In other words, they had multiple reasons to be afraid of my presence. It took about a year to earn their trust and dissolve their doubts about me being a spy whose goal would be to send them to prison again.

A different problem emerged, however, when their approach to taking advantage of my presence was pushed too far. I collected scrap metal and distributed my finds among the waste pickers. I kept changing the recipients to prevent feelings of injustice. From my perspective, the donation of my finds was a fair approach because I was already paid as an academic. I did not want to increase the number of people who would divide the scavenged finds among themselves. The more people, the less money. While waste pickers generally appreciated my approach, some of them took it as an opportunity to test how far they could get with my generosity and started asking me for money loans. I resisted, but once I decided to approach it as an experiment in which I might lose my money but also learn something new. Soon after loaning the money to one of my closest interlocutors, he moved with his family to a different region and, therefore, disappeared from the landfill. Despite my anticipation that something like that could happen, this event taught me that keeping a certain distance is useful. Also, it taught me to evaluate carefully what my interlocutors said. Some of them were skilled in establishing what Antonius Robben (1995) calls *ethnographic seduction*. By overwhelming me with emotions concerning their impoverished lives and hardships, it was challenging to resist their stories about victimization. As Robben argues, victims are not devoid of the capacity to manipulate the researcher, who needs to think critically not only about the narratives of the victimizers—or those in power, broadly speaking—but also about the testimonies of the victims. Although seduction is not easy to discern, I applied two strategies to cope with it. Whenever possible, I triangulated the stories and claims from multiple sources and repeated the same questions a few times, separated by time gaps to reveal any inconsistencies.

* * *

The initial quotation from Gregory Bateson's work set a tone for reflexive thinking about how anthropological knowledge emerges. I described my transition from archaeology to ethnography and explained what it entailed. Also, I suggested that fluidity can be an epistemic tool for research. I never pushed my interlocutors to take care of me. I rather waited for opportunities that the waste flows generated. When I could, I simply stepped into the current and let myself flow with other people and materials.

My research among actors in waste management increased my sense of humility. It is striking how much I still do not understand. Time and patience helped me to decode some of the puzzles, inconsistencies, or contradictions, but there are still gaps in my knowledge. Some of those gaps were clearly influenced by the patchy ethnography and the fact that most of my experience relates to the work environment. David Zeitlyn (2009) understands incompleteness as an inherent part of *merological anthropology*: the project that acknowledges the partiality of anthropological views. Such views inevitably capture only parts of the world in the process of becoming (see Ingold 2013). While my research shows limits in terms of knowing all the diverse relations, actors, and spaces associated with waste management in a specific country in Central Europe, the length of my interaction enables me to recognize the incomplete process of becoming. Not only am I describing the changing waste regime during the last decade but, even as I describe it, I am already discerning new impulses that invite further continuation and collaboration.

NOTES

1. My studies, both in Czechia and the US, were strongly shaped by "four-field" thinking about anthropology. In all my theses, including my Ph.D. dissertation, I tried to use and combine knowledge, skills, and perspectives from archaeology, sociocultural, and biological anthropology. Anthropological archaeology, nonetheless, was the discipline that shaped primarily what and how I was conducting research even several years after my dissertation.
2. Later, we elaborated on this topic in our handbook of research methods in waste studies (Sosna and Brunclíková 2015).
3. I use pseudonyms for the names of places, companies, and people throughout the book unless stated otherwise. The Pureland landfill became the key site for my early ethnographic research.
4. These chimney-like metal objects are large pipes that serve to release landfill gases from the landfill's body. These gases are among the strongest generators of odor at landfills because they are products of decay deep in the body. Close proximity to a chimney can cause a sense of dizziness.
5. See also Fredericks, Rosalind. 2009. "Doing the Dirty work: The Cultural Politics of Garbage Collection in Dakar, Senegal." Ph.D. dissertation, Geography. University of California, Berkeley.
6. Light detection and ranging (LIDAR) is a remote sensing method that uses laser to generate images for geographic information systems. For the purpose of my research, LIDAR images of landfills enabled me to see their shape, size, and relation to their surroundings without vegetation. Since the laser can go through the vegetation, one can get a better sense of the topography of the ground surface.
7. Wilk, Richard R. 1981. "Agriculture, Ecology and Domestic Organization Among the Kekchi Maya." Ph.D. dissertation, Department of Anthropology. Tucson, University of Arizona.
8. Wilk (1981: 11).
9. Wilk (1981: 3).
10. I have never seen workers wearing face masks during interactions with leachate. In pre-COVID times, face masks were not even imagined as relevant, and the situation did not change much with COVID-19, which was not perceived as more dangerous than other hazards that could harm or kill one at the landfill. Moreover, wearing a face mask is highly challenging during heavy physical labor, which requires intensive breathing.
11. Czechia's lack of a colonial history and the specific development of anthropology in the country (see Balaš 2018) make practicing of the classic Malinowskian model of fieldwork unlikely. Neither academic institutions nor funding bodies support the absence of a researcher for prolonged periods of time. In contrast to the "patchy

approaches" developed elsewhere, which were strongly shaped by the COVID-19 pandemic, the back-and-forth model has much deeper roots in Czechoslovakia.

12. Along with their marginal spatial position, odor is one of the major reasons why waste pickers and workers do not use public transportation.
13. As I learned recently, local creative spirits developed vernacular labels for scholars in social sciences and humanities. The former scholars are called *socky* (welfare trash), the latter *humáči* (human filth). There is not much in these labels that would serve as a spark of excitement for the people working in waste management.
14. Interestingly, the capacity for concealment can be extended beyond the human to the species that became essential for my research: ravens. Ethological research revealed that ravens not only hide food in caches but also actively withhold information from other individuals (see Bugnyar and Kotrschal 2002).

CHAPTER 2

UNDERFLOWS OF WASTE STREAMS

> So I have been pressing old paper with a hydraulic press for thirty years, working hard in the cellar under the light bulbs, ... horns of abundance shower down from the sky, ... old butcher shop newspapers full of blood, wilted flower stems and bouquets of past birthdays and name-days, ... sometimes a newspaper-wrapped cobblestone, cardboard knives and scissors, ... funeral wreaths to make the paper weigh more ... through the hole in the floor of the yard above, events and stories from the past pour down on me and at my feet, ...
>
> —Bohumil Hrabal, *Loud Solitude* 1994: 85–86

It was in the spring of 2014 when the landfill worker Jindra and I were standing on a platform close to the edge of the active zone in the Pureland landfill, watching a garbage truck tipping its content. Since Jindra's boss, the landfill manager, was not around, Jindra opened the massive door of a six-meter-long and over two ton storage container (Figure 2.1) standing right next to us, attached a piece of metal sheet into the grip, and started grinding its edge while chatting with me. When I asked him about the origins of the container, which was normally used for the transport of commodities on board ships and trains, Jindra told me:

> Well, it is our container. We got it when we were moving from the original entrance to the landfill to the new one. We used to have a shed down there and we stored various stuff there. When they closed the original entrance and tore down our shed, we had to relocate our stuff. So, I got this container. It cost 30 thousand.[1] We simply managed to transport it to the landfill, so it is here right now.

I was surprised to hear that the container was not the company's property and that the landfill workers purchased it and brought it to the landfill. They invested a substantial amount of money to create their informal

Figure 2.1. The landfill workers' container for storage and tinkering. © Daniel Sosna.

infrastructure. Moreover, it was far from being a tiny marginal thing that could be overlooked. In contrast, this light-green container was located in the highest and most prominent place as a dominant sign of storage. It was literally the first thing that every truck driver saw when he or she approached the edge of the tipping zone and started descending to the kingdom of garbage. The recovered objects collected by the workers were stashed in the container. After their initial disappearance from the social contexts where they were found unwanted and subsequently disposed of at the landfill, they disappeared again behind the heavy door of the container. Moreover, this double-leaf door was equipped with a sturdy padlock situated inside a narrow circular metal sheet that prevented the padlock from being cut off by bolt cutters. As the landfill workers kept noting, one never knew when bogeymen (*bubáci*) might show up to steal their treasures.[2]

Over time, I realized that the container was central to workers' activities, and its internal structure reflected the major needs. The container was divided along its long axis into two parts that reflected two main functions: storage and tinkering. Storage enabled the objects to wait in a state of limbo for weeks, months, or years before someone needed them. For more than twenty years, the workers were collecting objects of potential value and storing them in their former shed and later in this large container. One of the container's walls was covered with shelves filled with highly diverse things ranging from snowboards and tennis rackets to plastic bottles with detergents, antique brass handles, and a series of electric tools. The container represented a key part of the infrastructure that shaped preservation and transformation of value (cf. Thiemann 2024).

Tinkering took place in the second part of the container, which served as a small workshop equipped with a workbench with a vice, various mechanical tools, electric drills, grinding machines, saws, and brushes, all of them coming from the landfill as the workers proudly emphasized. The container was connected to the company's electric grid, so the workers could take advantage of the already existing waste infrastructure without worrying about paying for the electricity. The position of the container was ideal for combining the official duties of the workers with their informal activities. The door was always open during working hours, and a worker who was tinkering inside or in the immediate surroundings of the container could easily oversee the active zone where the trucks tipped their loads. Therefore, the workers could react immediately and switch from tinkering, jump into a front-loader or walk and resolve the issue that required their attention, most frequently the trucks that got stuck in the mud and needed to be pulled out.

Both scavenging and tinkering represented a major part of workers' activities. Although their official duties entailed especially monitoring the nature of garbage tipped out, its distribution and crushing with the two front-loaders, navigating and helping truck drivers, spraying landfill leachate, and taking care of grass and bushes at the landfill, they were able to spend a substantial portion of their time at work trying to salvage value from the morass of garbage. Like Hanťa, the main character from Bohumil Hrabal's novel that I quoted at the beginning, the landfill workers felt a commitment to monitor the incoming garbage and recover things that seemed valuable. Both scavenging and tinkering depend on an ability to see potential where most people would not see it. As Reno (2016a: 124) says, while scavenging is primarily about the ability to perceive value where others do not, tinkering is more about its realization. Moreover, as Francisco Martínez points out, reparation, which is closely related to tinkering, carries a "symbolic charge that honours care over wasting" (2019: 12). Scavenging and tinkering are entangled, and both require experience, imagination, and skill. Removing wires wound around a discarded mattress immersed in wet landfill matter can become a more creative activity than tinkering with frequent finds such as discarded lawnmowers. Moreover, scavenging and tinkering are connected to diverse ways salvaged things follow when they embark on their further journeys. Landfill workers use them individually, share them, exchange them as gifts, or sell them as commodities. All these activities and emerging relations feed what Millar (2018: 9) calls *forms of living* that unfold along the waste streams. Millar's emphasis on life attempts to overcome the trouble of identifying where work begins and ends, which is a productive way of thinking about people who come into direct contact with waste at Czech landfills and beyond. Their activities resist the salient social need to make waste disappear. They enable the discarded matter to appear again regardless of the official rules and normative expectations. This chapter will show the commitment to not letting garbage go but to developing various ways to salvage it.

THE HISTORIES OF VIBRATING FEET

Scavenging is something between recreational fishing and mushroom picking. One needs to be patient like a person waiting for a bite. As landfill workers like to say: "Everything will eventually come to the landfill. It is just a matter of time." When one needs a tent for holiday travel, new windows or water tubes for one's house, it just requires patience and attention to incoming trucks and loads of waste. It would be foolish to go and purchase these things in a shop straight away. A perception of scavenging clearly differed from shopping (cf. Reno 2016a: 113) and the workers never used this kind of association. It was not driven by a desire to buy or seen as an alternative to shopping but being driven by an opportunity and subsequent ability to take advantage of it.

The mushroom picking dimension stems from the embodied quality of scavenging. The workers always walk with their heads bent downwards and continuously scan the ground for the presence of scrap metal or other potentially valuable pieces of garbage. It is reminiscent of walking through the woods when a mushroom picker looks around, lifts the branches, and occasionally stops to inspect the ground. One never knows what can appear in garbage and one has to be prepared to dig into it. Gloves stuck into the back pockets of the working pants with the fingers out complete the picture of the scavengers always ready to engage with the filth.

For the workers at the Pureland landfill, this activity certainly was not a mere survival strategy (see Millar 2018: 12). All the workers expressed in various ways that working at a landfill was special. It could carry a stigma when, for example, one got hurt at work, and an ambulance transported the injured person to the hospital, where the medical staff was not particularly excited with the odors evaporating from the dirty apparel. A story of Karel, the most senior worker, and his deep shame at the hospital were voiced multiple times. Also, unknown consequences of a long-term exposure to the landfill full of decay, flying droplets of leachate, insects, and unexpected chemical substances was a concern. These drawbacks, however, were outweighed by advantages associated with the opportunities afforded by garbage. A sense of excitement comparable to mushroom picking and recreational fishing was omnipresent and culminated during the days when supermarket chains or home improvement stores sent their trucks to get rid of the goods that were considered unsellable. In these moments, the slow-moving tilted bodies of the "mushroom pickers" changed into athletes who rushed to the fresh piles to check whether anything interesting had arrived.

Scavenging is much more an exciting activity than labor (cf. Tsing 2015: 77–78). One never knows what may arrive at the landfill. One needs to be patient like a recreational fisher because there is always a possibility that an exceptional fish may swim somewhere in the waste stream. Like fishing with spoons, one cannot simply wait but one needs to walk along the stream, keep casting a spoon to test whether there is anything below the surface while imagining what a catch might be there. These *futural orientations* (see Bryant and Knight 2019) based on omnipresent anticipation and hope are pleasurable. Therefore, it is not only assembling and disassembling discarded things that generate satisfaction (Sanchez 2020) but also the search is pleasurable and satisfactory. Although the workers at the Pureland landfill expressed their fondness for omnipresent search many times, Jindra's imagination went further. Since he was close to retirement, he hoped for something really big that would finally change his life after more than two decades at the landfill. His hope was tied to an image of a large suitcase full of money disposed of in a container in an emergency by the Vietnamese mafia.[3] Every time we heard Jindra's wish, we all laughed and Jindra was laughing as well. At the same time, his eyes were full of excitement with a tiny flame of hope that even such a ridiculous wish is not entirely impossible.[4] Jindra has seen a lot during his career and knew that *anything could happen* at the landfill. One just needs to be patient!

* * *

Located in West Bohemia, the Pureland landfill is one of the large Czech landfills designed for the deposition of primarily municipal solid waste. It is located in the area of a former sand quarry that was founded in 1954 to supply surrounding villages and the city of Pilsen with sand for construction purposes. During the 1980s, the quarry was slowly transformed into a place for dumping waste. In 1994, a new sanitary section of the landfill was opened to fulfill the requirements specified by the Waste Management Act of 1991. The current owner of the land, the Voluntary Confederation of Municipalities, leased the landfill to the company Warm in 2004, which ran the landfill till its shutdown in 2017.[5] After the official shutdown for municipal solid waste, the landfill has served for the disposal of the solid incineration residues produced by an incinerator run by Warm.

The landfill is situated on the slope of a broad hill. The part of the hill previously eaten by quarrying was being filled with garbage from the bottom. During my research, the landfill was reminiscent of a volcano. When one entered the site and passed the fences, there was a gate with the offices located in metal cabins, and a small booth for a security guard. Right behind the gate, there was a scale for weighing the trucks. Then, a paved road wound toward the top of the hill, where one passed a fenced area for composting and reached an edge of the active zone. Workers' containers and another cabin were right at the edge of what was reminiscent of a volcano's crater below. The paved road ended with several concrete panels to give way to the world of rails in dirt, whose nature was changing with weather conditions. The crater the size of a football field was always in a state of dynamic change. Various piles were growing and disappearing under the wheels of the front-loader; a level was growing irregularly here and there, and deep, muddy puddles were forming and giving way to the hard, dusty surfaces during dry days. The position at the top of the hill made the space vulnerable to weather conditions. There were some trees in the north, but other parts of the landfill did not have any natural barrier that would protect it from the powers of nature, especially the wind, rain, and sunshine.

Although the place was changing quickly, a feature was constant. Trucks and loaders always created a sense of *vibrating feet*. Being a landfill worker means that one gets accustomed to this shaking condition. This seeming detail, however, saves lives. Pedestrians, including landfill workers, waste pickers, and truck drivers taking breaks, face significant risks at landfills (see Millar 2018: 64; Reno 2016a: 3). One can get run over by loaders or trucks easily, especially when they are reversing. I realized this when Petr, the front-loader driver, let me drive the massive machine. There is a blind spot behind the loader, which increases the risk of reversing over a person with the machine designed for crushing garbage, with its metal wheels the size of an adult person. It can easily happen, so feeling the vibration of the ground helps to recognize that something big is coming, especially during foggy and noisy days when one can easily lose a sense of where the danger is coming from.

During my research at the Pureland landfill, there were four landfill workers, three officers who alternated at the gate and monitored the incoming trucks, and the manager. The four landfill workers had been employed there for years because the work environment was stable. When one started, the perks of the job were so convenient that none of the workers seriously considered leaving. The most senior, Karel, had stayed for twenty-seven years, while the most junior, Slávek, had been there for eleven years. The four came from various backgrounds—Karel used to work in an agricultural cooperative, Jindra worked as a plumber and petty entrepreneur in vegetables, Petr used to be a carpenter, and Slávek an army officer. In the end, they all firmly settled at the landfill. Jindra and Petr were alternating as drivers of the front-loaders at the site, while Karel and Slávek were doing only manual labor.

The manager came to the landfill relatively recently in 2004 when his company, Warm, rented it. Previously, he worked as a manager of buildings in the heating plant, which was also run by Warm. Rumors say that the top managers in the company wanted to get rid of him, so they sent him far from the city to the landfill where he could do no harm. Whatever the truth behind the rumors, he retained ties to the heating plant. He had two offices, one at the gate to the landfill and the second in the heating plant downtown. His regime of movement between both offices during the week enabled him to oversee the landfill partially and participate in meetings at the headquarters as well.

To understand the workers' perception of the waste regime (Gille 2007), it is useful to review the changes in the flow of garbage into the landfill. From 2004 until 2015, the influx of garbage was continuously declining (Table 2.1). In just over a decade, the weight of garbage dropped by more than 60 percent. During the last two years of the landfill's functioning, the decrease was even more dramatic (1,147 tons in 2017) because a new incinerator was opened close to the landfill.

The minimization of landfilling is predominantly viewed as a positive trend toward a more sustainable future (e.g., EU 2013; Pichtel 2014: 16). This contrasts strongly with a perception held by the landfill workers whose view was unanimously negative: there was not enough waste! In fact, one of the most pervasive themes expressed by the landfill workers was nostalgia for the times when "one didn't need money" because everything could be found at the landfill. The following description by Petr, an operator of front-loaders, is instructive:

> Back in the days when we were located down the hill next to the old scale, we had extra space for the storage of everything we wanted to keep. There was a garage where we even kept a couple of freezers to store frozen chicken and pork chops that were dumped in the landfill frequently. Carrefour [supermarket chain] was great. They

Table 2.1. Changes in weight (thousands of tons) of solid waste disposed of per year in the Pureland landfill. Source: Warm.

	Year											
	2004	**2005**	**2006**	**2007**	**2008**	**2009**	**2010**	**2011**	**2012**	**2013**	**2014**	**2015**
Weight	70.1	51.3	49.9	49.7	46.7	37.9	32.0	29.4	25.5	26.5	28.8	24.3

> were getting rid of large quantities of food that was still OK. We just took it and put it into the freezers. We had a little kitchen down there as well. Man, we were cooking every day. Our previous boss—it was a good guy, not like this one [i.e., current landfill manager]—came in the morning and asked "So, what we gonna eat today?" And we were cooking from the early morning. Golden times ...

Workers frequently recalled golden times of the early post-socialist days. Various companies were emerging and collapsing, therefore producing waste in the early days of the wild capitalism after the Velvet Revolution. During the 1990s, there were no official scrapyards collecting problematic household items such as fridges and washing machines as well as bulky waste (*objemný odpad*) such as furniture. This had two effects. The amount of garbage going to landfills was slowly declining with the development of scrapyards and incinerators. This is confirmed by the general trend of the gradual decline of landfilled municipal solid waste in the country up to 2015, as recorded by the Czech Environmental Information Agency (CENIA).[6] Also, the valuable items containing metals and engines diverged from the stream flowing to the landfills.

I still recall one morning in June 2016 when I came to the landfill and found Petr in a bad mood, standing on the platform next to their container. When I asked what was going on, he was silent for a moment, watching the tipping zone. Then he turned to me and said: "You can see how things are here. Look at it." He pointed his finger to the empty tipping zone. There was no truck. No piles of fresh garbage as usual. The silence was quite telling. When I looked around and saw half-empty barrels for the storage of different kinds of scrap metal, it was clear. During the whole day there were about ten garbage trucks arriving. No wonder that the workers were nostalgic about the past.

Workers' nostalgia was associated with the ideas of predictability and abundance, which were well comparable to other examples from Central and Eastern Europe. However, as Hann (2014: 117) argues, nostalgia in the region is a complex phenomenon whose nature may vary across countries. Among the workers in the Pureland landfill, nostalgia was not necessarily directed toward socialist times. Workers' references to "times of abundance" (*doba, kdy všeho bylo dost*) dissolve the categories of the socialist, post-socialist, and capitalist as irrelevant. It rather refers to the time when concerns about waste were limited and landfills were seen as natural vehicles for the disappearance of garbage from the public eye.

The workers marveled at times before the emergence of new policies responsible for the creation of new municipal scrapyards, with emphasis on recycling and incineration, which had been slowly transforming waste management in Czechia since 2001 when a new waste act was passed.

Their classification of the past entailed the good old time lacking public interest in waste—therefore, a lot of freedom for the waste workers—and the recent time when waste started inhabiting public discourses, environmental initiatives, and state policies, with the negative consequences for landfill workers. Back in the day, garbage was abundant and rules were more relaxed.

A specific position within the waste streams afforded the landfill workers a reduced perception of troubling or even tragic events of the past. For example, in 2002 there was massive flooding in West Bohemia. Despite being tragic for most people in the region, it had a positive consequence for the landfill workers. Both local entrepreneurs and large companies were transporting large amounts of waste to the Pureland landfill. Sometimes, the only problem with the goods from the shops was the absence of labels, which had become detached, and the surface not being clean. For the landfill workers used to scavenging impure things, goods from flooded shops were a real bonanza. They recall especially a truck full of wine bottles with damaged labels that came from a local liqueur store. The workers stored the wet boxes with bottles in their large garage, consumed alcohol together for several months, and smuggled some of the bottles out of the landfill for their relatives and friends.

Similar memories were associated with large quantities of coffee wafers, toothpaste, canned food, and cigarettes. Salvaging cigarettes, however, required special skills because they were brought by customs officers who used disposal in the landfill for the destruction of the confiscated goods. Interestingly, the strategy of the Czech workers was almost identical to that described by Reno (2016a: 114) in Michigan. They created a hole in the landfill, stuffed the bottom with soft material, deposited a load of cigarettes inside, and covered everything with a thin layer of garbage, having the officers behind their backs watching the operation. Once the officers had left, the workers carefully scraped the garbage off and retrieved the contents of the hole. The officers, who were not used to the dirt and odor of the landfill, probably could not imagine that those civilized guys in clean working apparel sitting in the front-loaders would be willing to dig out the cigarettes buried in filth. Of course, they could, because it was their everyday activity! Long-term direct exposure to garbage sets different limits for disgust (see Butt 2023: 129; Millar 2018: 61).

The workers always felt proud of outsmarting the officers or developing ways to take advantage of the discarded matter that others disposed of and assumed that it was the end of the story. Workers' pride is evidenced in their roguish grins and excitement while narrating significant past events and in their creativity in altering the official waste streams. It was a way to articulate their special position vis-à-vis flows of value and to emphasize their position as guardians of these flows. These escapes to the past reflected declining amounts and quality of garbage that was coming to their landfill as well as shifts in policies and strategies of customs officers, who moved to the incinerators to make sure that the discards really could not be recovered.[7]

UNDERFLOWS ERODING THE FORMAL VS. INFORMAL DIVIDE

Work performed for wage rarely sets a boundary for conduct that could not be extended beyond it. It is usually about the degree of the boundary's porosity or fuzziness. The world I describe in this chapter falls close to the end of the spectrum where formal boundaries become fuzzy. Workers' scavenging and tinkering were intertwined with the official requirements of their jobs, which primarily entailed the monitoring and distribution of waste within the landfill. This work ethic used to be common among the workers during state socialism (Wedel 1992: 76). I recall factory workers from my childhood who were able to produce various things during their "free time" at work. When one needed a metal construction for the kitchen desk or weights for weightlifting, they could be made at work and then smuggled home. Landfill workers at the Pureland landfill combined their official duties with their informal activities easily. For example, on the way to pull a garbage truck from the mud, the workers walked and examined the ground and piles for potential finds. Once they transported the finds to their working area near the container, they either started processing or repairing the things immediately or later when there was nothing else to do.

Since the workers had to repair their working tools or components of their front-loaders from time to time, it was hard to make a clear line between what could be understood as their formal and informal activities. For outsiders visiting the landfill, everything looked natural, as part of the workers' jobs. Indeed, these activities lacked any categorical distinction for the landfill manager and the workers themselves. They became so accustomed to scavenging and tinkering that they did not distinguish them from their official work. Prescribed versions of conduct were not strictly specified anyway, so there was always space for doing something else or fulfilling work tasks in ways that enabled their adjustment, plastic reshaping in Millar's sense (2018: 15). Formal work was surprisingly open-ended. Moreover, even the official duties were subject to negotiation. The manager did not want to lose the workers because it would be difficult to find new ones.[8] Therefore, he was careful to not criticize them for performing the activities not ordered by himself. The workers felt this hesitation and used it to maximize the extent of their own recovery activities.

There is an important point concerning workers' previous lives. The youngest one was born in 1968. This means that all of them were socialized into the working environment during the era of state socialism when informality flourished. Importantly, the manager had this experience as well. This may explain why the workers disregarded the difference between their official work and other activities at work. Also, it suggests why the manager, who was exposed to the socialist moral economy in his previous job in the heating plant, could understand such informality.

Informal economy has become a pervasive concept that sparked not only academic debates but also colonized the spheres way beyond them. When Keith Hart (1973) elaborated on this concept to theorize about labor and diversity of income opportunities he probably did not expect the journeys the concept would undertake in the next half a century. A specific understanding of informality emerged from studies of socialist and postsocialist societies, particularly from Ledeneva's (1998, 2011) and Humphrey's (2002) work on informal rules and practices. Informal rules are not articulated, yet actors know and follow them to satisfy the needs that cannot be satisfied through formal practices. Further studies showed high diversity of ways informality is understood, talked about, and lived (Grygar 2016; Ledeneva et al. 2018; Polese 2023), *favours*[9] as a pervasive feature of informal economic practices (Henig and Makovicky 2017), their scalar dynamic (Polese and Morris 2015), power relations (Stan 2012), and a role of informality in food production (Jehlička, Kostelecký, and Smith 2013).

The first critiques of informality emerged early. Jan Breman (1976) argued that the informal sector is a fuzzy concept, which is often defined through the absence of features such as a well-organized labor structure, legal protection, and registration in official statistics. The binary contrast between the formal and informal is rather artificial because formality and informality are aspects of the same social life and, therefore, mutually entangled (Bear 2013: 378; Guyer 2004: 165; Holmström 1984: 320). There might be good reasons to tolerate informality by states' apparatuses because of the advantages it can generate (Hartman 2008; Rubbers 2007).

There have been attempts to overcome the troubles associated with the concept of informal economy or informality. Ida Knudsen and Martin Frederiksen (2015) proposed the concept of *the grey zone*[10] as a tool for analyzing ambiguous phenomena that do not fit into neat dichotomies, including formal-informal opposition.[11] Others decided to shift attention from informality to closely related concepts such as favours (Henig and Makovicky 2017), corruption (Rubbers 2007), and illicitness (Gregson and Crang 2016). Although the critiques of informality have been substantial, there are also defenders of the concept who point to its heuristic usefulness to frame a general area of research interest and stimulate discussions among scholars from different parts of the world (Polese 2023). Using the concept of informality not as a target of analysis but as a kind of keyword for approaching a broad discourse community of scholars tends to be used widely. Such a pragmatic approach works well when the researchers focus on native conceptualization of informality as an ethnographic category (Makovicky 2018; O'Hare 2022a: 121) and its performative and affective qualities (Rakowski 2016).

Arguably, the most persuasive critique of informality in waste and discard studies was presented by Millar (2018), who argues that the problem of informality stems from its negative definition as formlessness. Millar recognizes form and structure generated by her *catadores* who collect waste at a Brazilian dump site and proposes a concept of plasticity to capture changes of form while allowing for the existence of the order. In her view, plasticity enables the capture of the dynamic of adaptation to new conditions via both giving and receiving form (Millar 2018: 127). At the same time, plasticity shifts attention from the purely economic way of thinking to what Millar calls *forms of living*. While I am not convinced entirely by Millar's shift to plasticity because of the divergent meanings of the term, I embrace her point about the ubiquity of the form. This direction of thinking resonates with Eduardo Kohn's (2013) theorizing about the form in a processual sense. For Kohn, form is not rigid stability. It is patterning with temporal quality. It lasts and propagates in the world. More importantly, form constraints possibilities, so the process of its emergence and propagation is necessarily tied to acts of restraining (Kohn 2013: 157). If we accept Millar's view about the ubiquity of form, we might pay attention to the emergence and propagation of configurations constraining possibilities across the formal/informal divide.

Yet there is another possibility to conceptualize different lives of form. Cleo Wölfle Hazard (2022) developed the concept of underflows to capture dynamism in hydro-ecology and extended it as a conceptual tool for thinking about diverse unseen flows. Wölfle Hazard's metaphorical thinking moves from the river water that seeps through gravel and soil to form underflows beneath the surface to the spheres of discourses and politics. Underflows serve to direct attention to "latent discourses that lie hidden or disregarded, like fluid underflows are hidden beneath a stream's bed" (Wölfle Hazard 2022: 8). To push this kind of thinking further, I recognize the fluidity of waste streams as "flows of different materials that have distinct properties and are headed for different destinations" (Reno 2015: 559). There are two analytical moments that deserve attention. First, the flows do not exist only on the surface where discards move through the conduits of waste infrastructures. There are invisible underflows accessible only to those who have intimate knowledge and experience with them. Second, matter does not have to be the only entity flowing in the waste streams. There are flows of ideas, signs, discourses, pollution, or harm that do not make it to the surface and remain hidden.

Thinking with flow fits well with Kohn's (2013: 157) proposal to conceptualize ecology and economy not as separate domains but rather as part of a life of form. Interestingly, one of Kohn's persuasive examples of how the form emerges and propagates itself in the world is based on the dynamic of water. While Kohn uses a powerful image of a whirlpool to describe the emergence and ephemerality of form, his main point is to emphasize that form emerges because there are constraints on possibilities (Kohn 2013: 20). Water may always flow otherwise! These other flows, however, also become shaped by constraints on the possible and generate regularities and habits afforded by the morphology of the riverbed, wind, or activities of humans. To extend this kind of imagination into the world of waste streams, one may identify and analyze the underflows in Wölfle Hazard's (2022) sense. Skillful peeling of copper cables with a knife, formation of heaps of scrap metal, containment of aluminum rods and scraps in barrels, regular transport of scrap metal, narratives of resistance, all this can be envisioned as regularity and habit emerging from the constraints on the infinite ways to peel, accumulate, contain, transport, or talk. These are forms emerging when the landfill underflows get separated from the mainstream. They can emerge and disappear like a vortex of the whirlpool.

Activities of the landfill workers can be productively theorized as engagements with various kinds of flows. The workers direct part of the incoming waste stream beneath the surface afforded by the existing waste infrastructure. Scavenging, tinkering, exchange, and smuggling divide the official waste stream and redirect the flow of matter to different directions where former garbage ends up as furniture in a worker's cabin, fodder for the carp in a private pond, kitchen utensils, or in a collection of a specific brand of socialist road bikes. Paradoxically, these underflows are not necessarily below the surface thought of as a border of physical space. A striking *meisterstück* (masterpiece) of the workers is that underflow is constituted by a *capacity for recognition* rather than the mere physical presence. In other words, many of the activities of the workers and their results are clearly perceptible. Pulling the wires from the heaps, carrying old stoves in a loader's bucket, barrels full of aluminum ledges, storage containers for scrap iron, all of this is visible but not recognized as having little to do with the official duties of the workers. It is an underflow that flows without being recognized by outsiders and without being reflected by the landfill employees themselves. These flows are as natural as flows of water getting through sand and gravel of riverbeds. The river embodied by the waste stream has its underflows that disappear from the surface, remain unseen, or reappear again when joining the mainstream.

Still, these redirections and rejoinders are inherent parts of the dynamic of flow. They are mutually constitutive and cannot be clearly separated. In this sense, underflow seems to be a better conceptual tool than informality because its entanglement with the surface flow goes without saying. Moreover, flow is a process in the first place. It affords for change. Therefore, the question of existing forms or structures and their absence is less relevant. Flow has the capacity to generate the form and propagate it.

Wölfle Hazard's (2022) thinking encourages one to recognize different modalities of underflows. For the case of landfill workers, flowing requires paying attention to value. When the workers scavenge, tinker, or exchange, they modify flows of value. To realize value, the workers mobilize knowledge, experience, and skill during their interaction with the waste material. As Waqas Butt (2020a: 4) demonstrates, value gets realized by different actors at different points during the process of waste management. Therefore, imagining value as flowing is another way to work with waste's indeterminacy (Alexander and Sanchez 2018; Henig 2019). Since there are different modalities of value that emerge at the landfill and beyond, underflows enable the capture of those modalities that would get lost in the dominant story of discarded materials being the means for generating financial value. There are examples of a wall clock recovered for its aesthetic value, hockey sticks recovered for their use value as tools for cleaning the wheels of the front-loaders, or bike components for their value as gifts for friends.

Underflows also capture the knowledge that gets ignored despite being useful. The workers are never asked for their experience during building or modifying waste infrastructures, which can result in failures. The designers of the infrastructures rarely pay attention to the knowledge and experience of the people who spent years or even decades living with garbage in extreme mud or snow, or who facilitate the movement of vehicles weighing more than 10 tons. A good example would be the construction of a new paved road to the landfill when the location of the entrance changed in 2013. The workers suggested from the beginning that the path was not selected well and included a steep slope, which might cause trouble to the overloaded trucks during winter. Nobody paid attention to their comments. When a few garbage trucks slid off the road during the first winter and drivers started complaining that the road was reminiscent of a bobsled track, landfill workers just commented that they had known exactly that this would happen. As Petr said: "Nobody bothers to ask us." Over the years, I witnessed multiple suggestions of the workers to improve waste management, but nobody cared much about their expert knowledge based on direct long-term experience with waste management. Unfortunately, such a knowledge underflow remained hidden below the surface of the official debates among the designers, builders, and managers of Warm.

MODES OF ENGAGEMENT

In autumn 2014, I was collecting scrap metal with Jindra, the driver of a front-loader. We found some iron tubes during our inspection of the fresh piles of garbage. We carried the tubes on our shoulders and headed toward a large pile of scrap metal situated at the edge of the tipping area. When we reached the pile, he told me to throw the tubes on this pile used by all the scavengers. Later, when we were carrying aluminum ledges, Jindra instructed me to put them inside one of the metal barrels used for the storage of collected scrap metal. Perplexed, I asked him whether the distinction between the barrels was based on materials deposited there. The answer was negative and Jindra explained to me that each landfill worker collects his own scrap aluminum, copper, brass, and steel. When I noted that we were throwing scrap iron on the large common pile, Jindra was silent for a few seconds and then told me:

> We collect iron together and we really do not care who collects what. We simply divide the money into three equal parts at the end. Other metals, that's a different story. Everybody collects them on their own. We just do it that way.

Figure 2.2. Scrap metal collected during "free time." © Daniel Sosna.

This contrast struck me and prompted me to think about the logic behind these different modes of collection and how calculative reason played out in social relations mediated by material things. Stephen Gudeman's (2008, 2016) model of the tension in economy was a thread that led me to think about the collection of things in the landfill. Gudeman argues that economy grows from the dialectical tension between two realms: mutuality of social relationships and self-interested rationality.[12] He describes the dialectical relationship between the two realms as the dynamic potentiality of absorbing, merging, or subsuming features of the other (Gudeman 2008: 14). This kind of dynamic could be observed at the Pureland landfill, where one might identify at least three different modes of workers' engagement with garbage during their scavenging and tinkering. These three modes are as follows: engagement with things, scrap iron, and other kinds of scrap metals.

The first mode of things was most diverse. Certain things in the garbage were seen as potentially valuable for different reasons. Pliers, knives, hammers, and brooms were shared among the workers in the sense described by Thomas Widlok (2016: 87), who emphasized shared demands that fit neither the logic of reciprocal obligations nor commodity exchange. These things were used as working tools and became a natural part of the work environment. For example, when a worker found a hammer or a hockey stick, he used it himself, but others could use it as well without any obligation. When a work task required a tool, a person who needed it simply had an opportunity to use it regardless of who was the original finder. Such a sharing generally happened in an atmosphere of mutuality growing from workers' participation in the same work projects or sharing together the same destiny as stewards of the waste streams.

Sharing, nonetheless, did not relate only to tools used for labor. A few meters from the workers' storage container there was a construction trailer on wheels.[13] It was an old version of a trailer used during socialism by various field workers, miners, or scientists, simply all those who needed protection from bad weather or to sleep somewhere overnight. Workers had their official workers' cabin right next to the gate but it was located 300 meters away.[14] Having the trailer next to their container and the active zone enabled them to control the situation and react quickly to changing weather conditions or manager's orders. It included not only pieces valued for their primary utility, such as a table or radio but also diverse small pendants or an ashtray salvaged from the garbage for their aesthetic value. Electric heating and a sofa lent the trailer a sense of comfort. The cozy atmosphere was reinforced by the red wall clock featuring the Marlboro brand and an indispensable mirror for checking workers' hairstyles. It goes without saying that all the equipment was scavenged from the landfill. Former garbage was transformed to create a communal space where we could warm up during winter or heavy rains, chat, or listen to the radio. Recovered things were valued for their capacity to create a shelter and shared space for dwelling. The things were not moving. They rather became inherent to the space, a kind of inalienable possession that people never exchange (see Godelier 1999; Weiner 1992). Given workers' intensive engagement with various kinds of exchange derived from the everyday flow of fresh garbage (see below), these things represented an "effort to subvert change" (Weiner 1992: 11). They were material anchors of the workers' struggle to face the changing waste streams. In the world of a permanent flow of things and uncertainty about their amount and quality, the trailer represented stability.

When recovered things could not be used on the site, or if there were already plenty of them, there were other routes to salvage them. Things such as sporting equipment, electronic devices, books, or tools were given as gifts to relatives, friends, or acquaintances. Discarded packed food that was still edible was either consumed on-site or taken home and was often given as gifts to relatives or friends.

In the spring of 2015, Jindra found a well-preserved electric lawnmower. We transported it to the workers' storage and tinkering container. We spent the afternoon taking it apart, folding the plastic and metal parts, sharpening the blade, and playing with the electric wires. Finally, the lawnmower started working, and Jindra was happy to have a new item for exchange. Since he had already brought two much stronger lawnmowers from the landfill for his garden in the past, his thoughts were directed toward the network of his relatives and friends who would need such a device. A week later, he told me about his close friend who needed a small lawnmower for his garden. Jindra gave the device to his friend in exchange for future services. Jindra's friend occasionally helps him at home when the work requires more than one person. Moreover, he runs a small bicycle shop and repairs Jindra's bikes for free whenever needed.

Although gift-giving was the most natural way to use things scavenged at the landfill, the interaction could also follow a quasi-commodified logic (see Alexander 2009: 236). The price depended on the receiver's generosity. The workers maintained the attitude that "they will give me what they will give me" (*dá mi co mi dá*), but the transaction would be thought of in monetary terms. These exchanges demonstrate that a strict separation between gifts and commodities is not always possible (Appadurai 1986: 13; Miller 2001; Parry and Bloch 1989: 8–9; Stehlíková 2021). It is rather a continuum; some practices incline more toward the gift, while others are closer to the commodity logic of exchange.

The second mode concerned scrap iron. The landfill workers continuously scanned the surface of the landfill for the presence of metal bars, pipes, wires, or devices with a large proportion of metal parts. Iron was the most common metal present, and it was found in large quantities. Normally, the workers kept one or two large piles of scrap iron in the tipping zone where they accumulated their finds. Once in a while, they moved the contents of the piles to a large open storage container, and when the container became full, Jindra called his old friend in a scrapyard to arrive and pick up their container with iron. Once Jindra received cash, he divided the money into three equal parts and gave two parts to Petr and Slávek. The emphasis on the generous disregard of the contribution of the actors and equal division of financial gain gave a strong sense of mutuality.

The last worker, Karel, was excluded from this collective effort. He refused to contribute financially when the workers collected money to buy their own metal storage container. Therefore, Karel's position was somewhat ambivalent. Others were not happy that he refused to share the burden of their initial investment. Still, every now and then, Karel dropped iron objects on the common pile. He would not be able to transport large quantities of scrap iron on his own, so he followed a sense of mutuality in Gudeman's sense without calculating the consequences of his actions. Others understood these as acts not costing Karel much and still fitting into the shared sentiment that salvaging things from being lost was good.

The last mode concerned more precious metals such as copper, aluminum, bronze, brass, zinc, and lead. Although both iron and other metals were collected because of their exchange value as commodities, the difference in valuation shifted the emphasis on mutuality to a more calculative and individualized kind of reasoning, resulting in the storage of such finds in personalized barrels. During my research, one could get from 1 to 3 CZK (ca. 0.04–0.12 EUR) per kg of iron. The price of copper, however, ranged from 40 to 160 CZK (ca. 1.58–6.30 EUR) per kg. Moreover, more precious metals often required more labor. Copper wires and cables are often covered with plastic or located inside various devices from which they must be removed. Aluminum often appears in the form of profile ledges or tubes that have to be extracted from composite objects such as shower screens or furniture. All these reasons contributed to the engagement with these materials in a more calculative sense.

The process of disassembling requires knowledge and skill. It is a creative process that can be thought of as "craft working backwards" (Sanchez 2020: 76). Similarly to craft work, it takes time to learn it. Petr taught me how to separate aluminum and plastic parts of showers, but I never achieved his expertise. In fact, I became a subject of laughter multiple times because of my clumsiness. In the past, the workers could use less-demanding techniques. I was told that during the 1990s they used to burn copper cables and wires to remove the plastic covering. It was the time when discarded shoes were used as fuel for a stove located in the cabin. Toxic smoke coming from the chimney and open fires on the edge of the landfill to remove the plastic from the copper cords were no longer acceptable when I did my research. Therefore, the workers either had to peel cables and cords mechanically or deliver them to the scrapyard with the plastic covering still in place, which resulted in lower prices.

The collection of scrap metal is generally less enjoyable than the collection of other things (see Reno 2009: 42). Salvaging things that have their own "identity," hence they can be utilized or repaired, is rewarding because one can easily imagine the good that these things can generate as things. Anonymous metal bars, wires, or tubes that are often cut into smaller pieces or deformed lack the potential charisma of things. Their transformation via melting and other processing in unknown places somewhere far away does not afford the same affective attachment. Moreover, manual peeling of copper cords and cables with a knife is a toil. One of the workers, Jindra, even refused to engage in the unpleasant peeling despite the financial value of the raw copper. Being a few years from retirement, he ignored the cords and cables or directed others to go, recover, and process these findings. Jindra simply felt tired of demanding and dangerous peeling and focused on other materials and things.

The success in the collection of scrap metal varied but, in general, the workers could earn about a quarter of their monthly income via the scrap metal economy, which was a significant contribution to their salary.[15] To put the Czech situation into a broader context, it is important to emphasize that metals are the only materials worth paying attention to during scavenging at landfills. Ethnographic literature often describes people whose work is tied to the collection of a greater diversity of materials such as plastics and paper (Butt 2023; Doherty 2022; Millar 2018; O'Hare 2022b).[16] Such activities would be hardly imaginable in Czechia. I still remember the perplexed faces of my fellow landfill workers—and later the faces of waste pickers (Chapter 4)—when I mentioned that in some countries people collected plastic bottles to make money. It was hard for them to believe. In Czechia, this would not make any sense. The system of separating the collection of plastics into special containers is so developed that the main trouble is not the availability of plastics for recycling but the capacity and economic incentives to recycle the separated plastics. There is so much already separated plastic waste frequently ending up in incinerators or even landfills that the very idea of plastic collection among the scavengers sounds ridiculous. The abundance of discarded devices and scrap metal is so great that scavengers can be selective and focus only on the most valuable materials (copper), materials whose ubiquity can generate great weight (iron), or things that have aesthetic value, can be reused, shared, or exchanged.

ETHICAL FRICTIONS

Scavenging generally happens in a friendly atmosphere, and there is little sense of competition. Workers did not try to outrun others when a new truck tipped off the garbage. The rule of thumb was that a finder was considered the owner and could decide what to do with a find. Collaborations were common. When somebody found, for example, tiles or bricks and another worker was building a garage at home, the finder told the other person about the location of the find and even helped with the transportation. Individual workers had their own preferences, which were known by their colleagues. For example, Jindra owned a small pond with carp and collected old bread to feed them. Also, he collected old bike components for his friend, and when a component appeared, other workers brought it to Jindra. Slávek collected almost anything imaginable, but he especially liked to collect kitchen utensils for his wife. Since utensils from the landfill did not excite the wives of other workers, they usually drew Slávek's attention to such finds. Large things were not carried by hand but transported inside the front shovel of the loader. Therefore, the two drivers, Petr and Jindra, always had to help with the transport. They felt an obligation to help their colleagues, who acknowledged it.

An engagement of the workers with the underflows of garbage was generally perceived as a natural response to the conditions of abundance brought by the waste stream. I never encountered any sign of shame among the workers, criticism coming from the officers at the gate, or the manager himself. These activities were perceived as a perk of the job, not a theft (see Alexander 2009: 236; Pawlik 1992: 79–81). The workers talked without hesitation about their personal preferences in scavenging, the kinds of recovered objects their relatives and friends liked or did not like, projects that they had accomplished using materials from the landfill, or how they outsmarted customs officers with their fictive destructions of confiscated goods. Moreover, the workers enjoyed the disruptive potential of scavenging. As Reno argued, scavenging and associated activities "serve to disrupt idealized relations between landfill, home, and market" (2016a: 112). The Pureland landfill was a vehicle for turning things upside down: enjoying the landfill's abundance as a means to thrive instead of being faceless disposable workers dependent on the company's will.

It would be a simplification, nonetheless, to draw a picture of collaboration and mutuality accompanying every step of the employees at the Pureland landfill. The most senior worker, Karel, had a special position among the others. He was excluded from the collection of scrap metal because of his refusal to contribute financially to buy the storage container. Also, he provoked others with his careless approach to purifying practices; others often complained about his body odor and the smell of his clothes and boots. Although all the workers adapted to the landfill environment, Karel came the farthest after 27 years in the landfill. With a slight exaggeration, he became the landfill. His dirty, long hair with elf locks, droopy mustache, damaged, unwashed clothes, and slow motion was unsettling. He did not use gloves at work, which contrasted with other workers. He was the only one who consumed discarded food directly in the tipping zone. Similarly to purifying practices of separation between "work" and "home" described by Reno (2009: 40), the other three workers invested a lot of energy into cleaning their hands, changing their clothes frequently, and even keeping the toilet, shower, and the small workshop inside the storage container strictly clean. Karel did not fit this framework because he ignored most of these practices. Others forced him to spray his boots with recovered deodorants or *eau de toilettes* or even sprayed his hanging clothes in the cabin with deodorants themselves to fight Karel's carelessness in relation to odors.

When some official work tasks required collaboration among the workers, they did collaborate, and even Karel was part of the collaborative work. The workers respected Karel's extensive experience and ability to find things inside the piles of garbage. When the workers negotiated with the manager, occasionally resisted the orders, or complained about unjust treatment, their inclusion of Karel was strongly dependent on the context. It is important to note that Karel was the manager's favorite, who was willing to respect the orders and accomplish the tasks regardless of how dangerous or dirty the work became. It was hardly surprising that he was the only worker who was offered a new job by Warm once the landfill was shut down in 2017.

The landfill manager had an ambivalent relationship with his subordinates. He tolerated workers' scavenging because these activities had been tolerated long before he came to the landfill, and he saw value in these activities. His tolerance emerged from the perception that a good person should support this continuity, especially given the salvaging nature of these activities. He suggested a few times that he would feel bad for prohibiting such activities. This line of reasoning about the good was close to what David Henig and Anna Strhan (2022: 8) call the aspirational perspective of "*what ought to* or *could be.*" This perspective captures visions of possible alternatives to the existing structures and practices, which are not necessarily perceived as good ones. Also, scavenging was advantageous for the company because it increased the space available for the disposal of new garbage. At the same time, the manager disliked workers' frequent mumbling and complaints, related especially to spraying landfill leachate (Chapter 3). He felt that he was giving the workers enough freedom to scavenge and tinker to deserve their occasional resistance. They had a job, and he was their boss. Period.

It goes without saying that the workers' feelings were ambivalent as well. On the one hand, they acknowledged the manager's tolerance of their activities and considered him to be a fair (*férový*) person. In this context, *fair* did not necessarily mean *good*. It was an expression referring to his refusal to betray them or cheat on them. The manager was perceived as a person who did not care much about them, but he would not stab them in the back. On the other hand, they disliked his weakness, which had at least two different senses. First, it reflected the manager's inability to stand up for his subordinates and fight the system. The workers felt they were being treated with disregard for their work in the dirt and with unknown consequences for their health in the future. Their sense of injustice crystallized in a story that I heard multiple times over the years. They were invited to celebrate the company's anniversary at the headquarters of Warm. When they arrived at the venue, all the chairs were occupied and they could not sit down and have dinner with the others. The landfill manager did not take care of his subordinates and nobody else cared about them either. They were left standing in the large hall while others were either staring or completely ignoring them. The *humiliation* was so palpable that the workers just grabbed "a few schnitzels" and decided to leave for a nearby pub to have a few beers on their own.

This event was only the tip of the iceberg of the manager's lack of care, as seen from the workers' perspective. During the whole period of preparation for the landfill's shutdown, which took about three years, the manager did not express much interest in the workers' future. Nor was he willing to challenge any order coming from the headquarters. He simply passed the responsibility for conducting the work tasks to the workers and did not consider an option to negotiate with his superordinate in Warm.

The second notion of weakness was physical. The workers often ridiculed the manager for being a skinny person not adapted to the harsh landfill environment. They called him a tendon (*šlachoun*). When he visited the tipping zone during the cold months, he was always equipped with a warm jacket, scarf, and knitted hat. His outward manifestation of habitus contrasted strongly with those of the workers' stocky bodies, broad shoulders or large bellies. Workers rarely wore hats regardless of the weather conditions and they would never wear something like a scarf. Their necks were proudly exposed with the top shirt buttons undone. The wind played with the workers' hair, which reached a special quality in Petr's mullet inspired by young Jaromír Jágr.[17] A sense of masculinity was tied to the ways the workers' bodies faced weather conditions with pride and moved around. Their slow confident walk contrasted with the manager's quick stride and small steps. When they followed the manager and mumbled about their job, it was almost like a group of bears trying to keep up with a rabbit who was pelting along in front of them. A more pejorative nickname for the manager was Goebbels. Petr claimed that the manager was so reminiscent of the Nazi chancellor that he could not resist using this nickname. Although Petr claimed that it was the sheer physical resemblance that led him to create the nickname, it was clear that the word carried a large semiotic baggage. Nobody wanted to be called Goebbels in a country that had been occupied by the Nazis during World War II.

Moral and ethical intricacies became most evident in the case of scavenging scrap metal. This waste underflow depended on the ability of the workers to smuggle 6 to 10 tons of scrap metal out of the landfill every year. Strictly speaking, transporting already disposed waste out of the landfill is illegal. Czech law requires that a company running a landfill must prevent any theft of waste and its transport out of the landfill. Since the manager did not want to prohibit workers' activities but needed to minimize the risk of getting a fine from the Ministry of Environment, he created a special institution that the workers called a tithe (*desátek*). It was a payment that had to be made to their company to legalize the workers' scavenging.

The institution of the tithe comes from medieval times when villagers paid one-tenth of their annual produce or earnings to the Church. *Desátek*, paid by the workers, followed the same logic. Once per year, the workers had to officially announce and measure a load of scrap metal transported out of the landfill, so that the company would have evidence that the transport of scrap metal was organized by the company and, therefore, was legal. In other words, the workers and the manager presented it as an official treatment of scrap metal, pretending that this was all the scrap metal that had been collected throughout the year. The workers were convinced that the manager had not created *desátek* to keep the money in his pocket. All available information, including the manager's own oath, indicates that he really did report *desátek* to the authorities and knew that there was little danger of a detailed audit that would scrutinize the real scrap metal flow. Financially speaking, the amount concerned was ridiculously low for the company. The workers, however, hated *desátek*, and interpreted it as a kind of revenge by the manager for their occasional resistance to his orders. They argued that there were far more serious and legally threatening issues at the landfill, making their scavenging insignificant.

The landfill workers felt they had the moral right to earn extra money through scrap metal collection and sale because they did not feel they were being treated fairly within the system. Their mediocre salaries did not reflect the dangerous and unpleasant nature of the work they performed. Yet, their endurance was never acknowledged or rewarded. The company's leading managers did not show much interest in the workers, who were considered merely as disposable tools of profit—a common feature of the companies in the region (see Delanová 2015; Lupták

Burzová 2019; Sosna and Brunclíková 2019). As Petr, one of the workers, described it: "Look how these masters (*páni*) cheat. And they're gonna tell us something? Come on ..." The workers' moral justifications for their right to perform scavenging and tinkering reflected the unequal relations and distribution of power. The transgression of rules did not represent an issue for the workers because the rules were viewed as a vehicle of oppression rather than justice. When laws clash with moral values, especially when these values relate to equality and human dignity, such laws do not have to be seen as worth respecting. Human activities can be simultaneously illegal and honest when they derive from real honest work (see Frederiksen 2015: 166; Wanner 2005: 523). During state socialism, some laws were considered *immoral*. Given the workers' ages (forty-five-plus), they all had experience with this perception of laws and frequently referred to the fact that not much had changed in this respect during the preceding three decades.

From the perspective of the manager, establishing and requiring *desátek* was, therefore, an ethical act that enabled him to find a balance between formal rules, whose significance increased with the expansion of the neoliberal logic, and the moral value of scavenging. It was a favour, a specific mode of acting that did not necessarily reflect the logic of obligation (Henig and Makovicky 2017; Humphrey 2012; Ledeneva 1998). To enable the collection, processing, and transport of scrap metal required multiple small, everyday acts; the manager had to decide the amount of time he would tolerate for the scavenging activities of his workers, when and how they could use the front-loaders, and what portion of the smuggled metal should be legalized and reported. It was an ongoing exercise of judgment reflecting specific circumstances and being able to compromise when necessary. For example, during a rainy busy day when the garbage trucks got stuck in the mud or on a day after a TV broadcast investigating the grey economy, the manager was angry at the workers and urged them to really *work*, but when there was a little buzz and he was in good mood, the workers could spend almost all their time at work scavenging and tinkering.

Deana Jovanović (2021) raised another interesting point related to the illicit activities associate with stripping or collection of metals. She argued that these activities do not necessarily have to go against those in power but can even legitimize and reinforce their power. The manager's willingness to be complicit with the collection and smuggling of scrap metal was a way to reinforce the hierarchy and signal that he was the actor in charge who could tolerate or ban these activities according to his will.

Understanding of these moral and ethical complexities benefits from a perspective that puts into its center the everyday. Although scholars such as Sahlins (1972: 199) previously admitted that morality is not about the normative rules but emerges from specific relations and situations, this Aristotelian view has been developed especially in studies of ordinary ethics (see Das 2015; Keane 2015; Lambek 2010b). As one of the main proponents of this school of thought argues, understanding ethics as emerging through action enables scholars to account for uncertainties and ambivalence of social encounters (Lambek 2010a: 7). Therefore, rather than identifying the limits of the ethical or searching for its definitions, the interest lies in understanding "… how any concepts, including ones through which we try to catch a sense of the ethical, emerge in the give and take of life …" (Das 2015: 66). This means that it is not only about social life viewed through the prism of aiming at and doing good but also about the doubts, conflicts, or ambiguities that this project entails.

Specific situations at the landfill required the ongoing judgment of multiple and sometimes incompatible or conflicting components of ethical reasoning and action. Ongoing judgment is part of everyday action; its temporal directionality can be prospective, immediate, or retrospective, and include others as well oneself (Lambek 2010c: 43). There were antipathies and individual interests among the actors shaping the underflows of garbage as well as assumptions about the good. The employees navigated through the ethical landscape, sometimes feeling they were doing the right thing despite transgressing the formal rules; other times, they certainly knew that they were driven by more selfish interests. Given the dynamic nature of ethical reasoning and action, it was hard to predict the outcome of action. There was always a choice, the basis of morality and ethics in the economy (see Carrier 2017), which was lurking around. The resulting picture of the *recovery ethics* was fluid. Similar to the flow of water under the ground, it was constantly in motion, touching the limits of its movement, and unexpectedly seeping through the cracks while moving in directions that were never entirely predictable.

* * *

I began this chapter with the horns of abundance showering down to the cellar of Bohumil Hrabal's main character, Hanťa, who contemplated the events and people who produced these horns of abundance. For Hanťa, discarded things were material triggers of meaning and imagination related primarily to the past or the world of ideas outside the time. Although the workers at the Pureland landfill liked to recall the past, their everyday life at work was more about recognizing the potential of the present to shape the future. The recognition of which find was worth paying attention to and the estimation of its value that might be realized in the future for a wide spectrum of actors were the key capacities of the landfill workers. They were driven by the desire to deny the fate assigned to the discards: disappearance. This denial stemmed from the need to prevent value from being lost via crushing and burial of garbage at the landfill. Diverse engagements with garbage enabled the workers to pull their "fish" from the waste stream to let them flow in new waters. Regardless of whether their new life was tied to having use value as tools at work or home, aesthetic or sentimental value to co-constitute a living space, exchange value to be given as gifts, quasi-commodities, or being a primary vehicle of extra monetary income, the salvaged things and materials were emerging, transforming, sometimes sleeping in a state of limbo in the containers to emerge and engage with the outer world again. They refused to disappear.

The ethics surrounding practices of scavenging and tinkering mimicked the unpredictable flows coming to the garbage. The flows consisting of waste matter, trucks, people, rules, orders, rumors, and worries were flowing but one could never know what exactly would come and in what quantities. The relations among the actors carried a similar sense of unpredictability. Although the general values that salvaging things and helping each other are good, individual affinities and calculations shape everyday interactions and decisions. It was not unusual to see inconsistencies in the ways the employees interacted with each other and engaged with the garbage. A sense of what was right, acceptable, or wrong was emerging and was modified during action. It seems that the dynamic of the waste stream encouraged the actors to be always open to modifications and adjustments in their actions.

Skillful peeling of copper cables with the knife, formation of heaps of scrap metal, containment of aluminum rods and scraps in barrels, regular transport of scrap metal, and narratives of resistance, all can be envisioned as regularity and habit emerging from the constraints on the infinite ways to peel, accumulate, contain, transport, or talk. These were the forms emerging when the landfill underflows became separated from the mainstream. These underflows were flowing beneath the surface of the official stream, but it does not mean that they were always invisible and separable from it. The workers made some things disappear in their containers but large piles of scrap metal as well as their activities of scavenging and tinkering were overt. Paradoxically, the invisibility of these piles and practices stemmed from the inability to be recognized as underflows. This tactic followed the logic of the Czech proverb that "there is a shade underneath the lamp," which means that invisibility may lurk at the very center of the device that sheds light. Also, the employees took advantage of the existing infrastructure, enriched it on their own, and integrated it in ways that would make the conceptualization of the formal vs. informal untenable. It rather demonstrated creativity in engaging with waste.

NOTES

1. 30,000 Czech crowns (CZK) is equivalent to ca. €1,200.
2. The word *bubáci* is a euphemism used to talk about the Roma. The usual vernacular meaning of the word refers to supernatural beings that scare children. *Bubáci* are imagined as beings who rise when darkness falls. Their association with darkness makes a logical connection with the Roma, whose darker skin is considered to be one of their key phenotypical features in local ethnoracial imaginaries. Also, the usual association of the Roma with stealing overnight goes without saying. It does not mean, however, that this word is widely used in Czechia in this sense. Despite being brought in the region with a strong Romani presence, I learned about this word as a label for the Roma for the first time at the Pureland landfill. Although the word reflects racial prejudices, it is still one of the less derogatory ones. Workers' vocabulary includes much stronger expressions such as *černé svině* (black swines) or *černé kurvy* (black whores).
3. While it is not entirely clear why Jindra picked the Vietnamese, it fits a general sense of xenophobia among the people who feel threatened by foreign migrants and ethnic minorities. The Vietnamese mafia is known to be well-organized and successful in generating high profits, especially via illegal drug production and trafficking. It fits a general stereotype of people who might have a coffer full of cash somewhere and try to get rid of it during a police raid.
4. Urban legends featuring various treasures are not uncommon among the people who come into direct contact with garbage. Elana Resnick (2025: 5) describes a story of her interlocutor Baka, who recalled her cousin finding gold in a discarded shoe.
5. Pureland landfill is run by a joint-stock company Warm, where 65 percent of shares are owned by a municipality and 35 percent by a private company.
6. Czech Environmental Information Agency publishes data and yearbooks. Retrieved 5 September 2025 from www.cenia.cz/publikace/statisticka-rocenka-zivotniho-prostredi-cr/.
7. Although my field experience with incineration is limited, I did conduct research in two Czech incinerators and know especially one of them quite well because one of the employees enabled me repeated visits over the years. Given the observed breakdowns, mistakes in design, various informal solutions to keep the plant running, emergency burning of animal tissues during the bird flu pandemic, and associated silences about sensitive topics, suggest that incineration is not necessarily more reliable in making things disappear because of its sophisticated design and power of the fire. It is the opposite. The more sophistication, the more space for creativity. Tipping waste into the concrete storage bunker does not make it inaccessible. Before waste enters a furnace, there are multiple steps when "things can happen." It can be lifted with a grapple, moved, mixed, or removed from crushers that may get stuck. Direct visual inspection of destruction can be more challenging than in the case of landfilling.
8. Czechia is known for its long-term low unemployment rates. While ethnographies of the workers from Central and Eastern Europe often examine the effects of harsh unemployment (Kideckel 2008; Rakowski 2016), the conditions during the period and in the region I describe have been rather opposite. Since 2013, the unemployment rate has been declining or fluctuating around 3 percent. For managers in the waste industry, it has been challenging to find new employees.

9. The word "favour" follows UK spelling as a reference to its use as a specific theoretical concept used in academic literature.
10. The word "grey" follows UK spelling as a reference to the original publication of Knudsen and Frederiksen (2015).
11. Interestingly, the term *gray zone* had already been used by Šiklová, Poláčková-Henley, and Turner (1990) twenty-five years earlier to describe something different: a silent group of people who were not resisting the Communist system in exchange for certain benefits. For these authors, the gray color was associated with dirt: "... the hands of these people will never by entirely clean, their moral credit entirely clean" (Šiklová et al. 1990: 351). Still, the gray color carries a sense of contested space of moral ambiguity, as in the case of literature on the informal economy.
12. Gudeman's thought has been developing from the earlier model of the tension between mutuality and trade to the less binary model where mutuality and self-interest get mobilized across the scalar registers.
13. Similar to the storage container, the trailer was brought to the landfill by the workers themselves. The previous owner wanted to get rid of it, so it was cheap to get it. Fortunately, the landfill manager did not object to the trailer's presence at the landfill.
14. Unwillingness to go 300 meters to the gate was not necessarily a sign of laziness. Entering the cabin required purification rituals to prevent its pollution. It was a place for eating and filling in paperwork from time to time. One had to leave the second skin (see Reno 2016a: 40) outside before entering. Taking off their boots, pants, and jackets was time-consuming. In contrast, the entrance to the trailer was less demanding. One was expected to remove only large pieces of mud from boots and apply salvaged deodorant on the boots or other bad-smelling parts of the garment to make the presence in the tiny trailer bearable. Using the trailer was simply practical for multiple reasons.
15. In 2016, the salary of a landfill worker was 23,000 CZK (850 EUR). The drivers of the front-loaders earned slightly more. These figures were close to the median salary in the country and at the upper limit of workers' salaries (CSO 2016).
16. See also Furniss, Philip J. 2012. "Metaphors of Waste: Several Ways of Seeing 'Development' and Cairo's Garbage Collectors." Ph.D. dissertation, Oxford, University of Oxford.
17. Jaromír Jágr is a phenomenal Czech hockey player famous for his curly mullet. He is the second most productive player in the history of the NHL after Wayne Gretzky and still plays the Czech national league at fifty-two years old (as at 2024). In Czechia, he is a legend of its own kind, hardly comparable to anybody else.

CHAPTER 3

CIRCULAR DISAPPEARANCE OF WASTEWATER

> ... recycling is not the logical outcome of an economic strategy but simply a way to keep matter circulating as long as possible, for want of a place where it can be permanently stored and disposed of.
>
> —Romain Garcier, *One Cycle to Bind Them All?* 2012: 77

A week before Christmas of 2015, I received an unexpected Christmas gift. It was a warm winter day without snow but with a lot of mud and occasional showers that did not make our scavenging at the Pureland landfill particularly pleasant. Around 10 am Jindra, one of the workers, and I were already wet and decided to go to the trailer to get warm. Once we settled inside, Slávek, another worker, entered the trailer in a bad mood. He had just learned that the company would have to reduce the number of workers because of the declining amount of garbage coming to the landfill. He knew well what this meant. As a former army officer and law student, Slávek had an excellent capacity to grasp the power relations and estimate the outcome of this news. His frequent absence from work because of health issues and tense relationship with the manager made him the most likely candidate to be sacrificed. He often argued with the manager, and there was a rumor that he had even attacked the manager physically in the past. Slávek kept complaining about the system for half an hour and then realized that he had a job to do. He was supposed to check an underground sump for landfill leachate because it had been raining for several days, and a capacity limit could be reached. Landfill leachate is a kind of wastewater that emerges through percolation of rainwater, from inherent water in solid waste, and through activities of microorganisms responsible for the decay (Teng et al. 2021: 1). When Slávek invited me to join him, I was surprised. From the beginning of my research, the manager had tried to keep me out of the sump, so I was excited that Slávek was willing to take me there.

We could walk, but Slávek decided to drive to the base of the hill where the facility for leachate management was located. We had to pass through a booth with the security officers to get there by road. Fortunately, they were in a good mood. Since I was as dirty as Slávek, my presence did not raise any suspicion. Indeed, one of the officers offered me a shot of slivovitz with a slice of bacon with a conspiratorial grin. He knew that one's stomach and mind needed to prepare to inspect the landfill's most lively internal structure. A combination of his smile and mustache, yellowed after years of apparently heavy smoking, was disarming. I took the shot, ate the bacon, and prepared to visit the sump.

First, we opened about an eight-meter-deep shaft reminiscent of a well. The reminiscence was only partial, because a dark brown, stinking liquid was inside. Slávek directed my attention to the lower rungs of the iron ladder leading to the liquid. They were narrower and looked fragile after years of exposure to the leachate's agency. Slávek told me that somebody had to climb down to do the cleaning from time to time, but nobody was eager to do so. An image of broken rungs and a fall into the dark depth with the leachate was terrifying. Still, it was only a prelude to what was coming next.

Slávek unlocked a nearby small house that led into an underground space. When we entered, the air was so humid and volatile that it was difficult to breathe. There was the underground system of concrete pools and metal pipes drowning in a dense fog through which we walked on shaky wooden stairs and bridges. Beneath our feet, a strong stream of leachate was rushing from the end of a massive pipe (Figure 3.1). Slávek claimed that the strong stream resulted from the mere gravity of the fluid being collected via a system of long pipes going through the bottom of the whole landfill. It was like something from a different world, hardly comparable to anything I had experienced before. We were inside the body of the landfill with its belly filled with decaying matter. The body was warm, and its interior was wet with fluids flowing through the vessels and tubes. One could feel that this digestive labor made the landfill sweat.

When we got out and could breathe freely again, I started getting the point about the manager's silence concerning the leachate sump. The quantity of leachate was massive, the current was strong, and the nature of the fluid was scary. Neither the workers nor I needed a detailed chemical analysis to know that this indeterminate fluid could *bite* (see Hird 2012). We could see its teeth's strength on the iron components of the underground infrastructure. Landfill leachate is arguably the most critical matter at sanitary landfills, which are designed to dispose of municipal solid waste. Leachate is a fluid found in the body of landfills inside the pipes that drain it and inside

Figure 3.1. The underground sump full of landfill leachate. © Daniel Sosna.

sumps or pools that hold it. It is entangled with the activities of bacteria that metabolize the waste matter to co-produce leachate. Leachate contains pollutants such as degradable organic matter, inorganic macro components, heavy metals, and xenobiotic organic compounds (Kjeldsen et al. 2002). The concentration of pollutants varies between the landfills (Teng et al. 2021: 10) and within a single landfill because of the different decomposition stages in its various parts (Kjeldsen et al. 2002: 298). As a consequence, leachate toxicity is difficult to predict (Hird 2012: 465). Its indeterminacy means that it does not automatically fall into the *hazardous waste* category; nonetheless, its toxicity can be high (Wynne 1987: 457). In addition, there is a risk of leaks when leachate gets through the plastic liner below the body of the landfill, which is the barrier that separates it from the surrounding environment. Even years after the landfill's closure, leachate can linger and spill.

Landfill leachate has a unique position in the story of waste's disappearance. While garbage may be transported away, buried, and therefore at least potentially hidden from human senses, leachate is more difficult to contain. Although engineers develop technological solutions to prevent leachate leaks, it is difficult to succeed because leaking is the primary capacity of leachate (Hird 2012: 458). It moves and seeps through layers of waste matter and gravel. Its fluidity and causticity enable it to leak through cracks in concrete pools or punctures in plastic lining. Moreover, its pungent stench can travel beyond the landfill to inform the outer world about the presence of decaying waste matter (Stamatopoulou-Robbins 2020: 64). This capacity of leachate to challenge containment confirms Michel Serres's (2011) claim that emanation and percolation inevitably spread pollution in space regardless of how much we try to contain it.

Containment of leachate in the sumps, pools, or cisterns, nonetheless, is only part of the struggle for waste's disappearance. There is a technoscientific solution called leachate recirculation (Knox, Beaven, and Cossu 2018). It is a process when leachate, after its collection in the sumps or pools, can be applied directly onto the landfill's surface or injected into its body using injection trenches or wells (Beaven and Knox 2018). Recirculation introduces a circular movement that brings the liquid back to the landfill, lets it percolate, gets collected, sprayed, or injected into the body again, and so forth. The process minimizes leachate treatment costs because transporting and processing it in a special treatment plant designed for hazardous waste processing is expensive.

In this chapter, I examine the practices and ideas associated with the recirculation of landfill leachate at the Pureland landfill to understand the relationship between the circle as an abstract ideal and the circle as an economic model. I explore what forms of representation the circle engenders and what consequences circular dynamics produce. I pay attention to the techniques of spatial dislocation and temporal postponement, as encouraged by Romain Garcier (2012) in his study of spent nuclear fuel. Using the case of nuclear waste management, Garcier demonstrates that circulation can be a means to keep toxic matter in motion across space and to postpone the problem of its final disposal. In addition, long-term circulation can conceal dangers resting in the toxicity of the circulating matter. Also, I draw on Alexander and O'Hare's (2023) ideas about *technologies of unknowing* to examine how landfill leachate dangers disappear.

CIRCULAR ECONOMY?

One of the ways to understand the recirculation of landfill leachate is to approach it as a kind of circular economy (CE). Recirculation is based on processes that organize matter, energy, technology, and labor using the logic of circular movement. Moreover, recirculation can be thought of as an economic process because the logic that drives it relies heavily on a need to minimize expenses necessary for leachate's capturing and neutralization. Although the promoters of CE might not, perhaps, conceptualize the recirculation of leachate as a typical example of CE, I encourage the reader to think about it as a kind of CE for two reasons. First, the process fits the general model of an economy organized according to the circular principle of waste prevention (cf. O'Hare and Rams 2024b: 2). The case of leachate is unique because it prevents a new kind of waste (liquid) from emerging from the matter that has already become waste (garbage). In a sense, it is a circular kind of prevention submerged into a larger circle of imperfect waste prevention that generates garbage as its unwanted residue. Second, several decades of experience with the leachate's circular movement generates productive feedback for understanding any circular economic process and its consequences.

CE has received attention since the 1970s when various scholars started building models of economic systems based on the idea of closed loops (Geissdoerfer et al. 2017: 759). An essential part of this project has been its emphasis on the restorative and regenerative nature of the economy. This is not entirely unknown in anthropology, where the ideas of regeneration resonate in studies of human perception and reactions to death (Bloch and Parry 1982). David Graeber's (2012) further elaboration on the concept of *cycle* emphasized the tension between stability and change: "The cycle is ultimately our way of imagining stasis, the steady state, the condition in which an object might be said to be in motion but remains nothing but itself, just as in each daily or seasonal cycle, the nature of each day or year can be seen as exactly the same as any other" (Graeber 2012: 280). Building upon the previous ideas about the regeneration of life, Graeber, nonetheless, makes an exciting move directed toward the economy.[1] Recycling and ideas about death and regeneration of life have something in common. They both face the problem of keeping the status quo while being exposed to omnipresent change.

CE seems to exploit the idea that circular movement is good because it is so natural. The circulatory systems of the body, such as the heart and pulmonary circulation, demonstrate that this kind of movement is essential for life. The same is true for the ovarian cycle, which provides the basis for the continuation of life beyond the limits of an individual, and various biochemical cycles, such as Krebs or Calvin cycles, that enable organisms to manage essential biochemical processes of their bodies. The natural cycles of water, nitrogen, and carbon provide similar inspiration for imagining resource management (Garcier 2012: 83). CE takes advantage of such a salient imagination related to circulation or cycling: the two processes based on the idea that the end is a new beginning. CE calls for an economy centered around closed loops and implies dynamism within overall stability. It thus constructs a seemingly new economic order while referring to the familiar examples from nature. While CE constructs nature as separate from human activities and a positive guide for economic schemes, leachate recirculation draws our attention to the seemingly separate natural processes entwined with a human-made world and its technologies. It suggests that the distinction between biological/technical parts of CE, which features in contemporary models of CE, is instead a pragmatic tool to increase the persuasive power of this kind of economy rather than a depiction of how circularity works (O'Hare and Rams 2024b: 8).

Such an entanglement resonates with Eduardo Kohn's (2013: 157–65) call to reconsider the boundaries between ecology (nature) and economy (culture). He proposes focusing on the regularities, habits, and patterns that are aspects of the form that emerges and propagates in the world. Such a process can orchestrate water, plants, and humans, as well as their economic and political relations. Conceptualizing the form this way enables one to consider entities of different kind as part of a single relational field that constitutes and alters the form. This view is particularly useful for examining the recirculation that mobilizes water, wind, evaporation, bacteria, workers, pumps and hoses, cisterns, and ideas about economization and harm to resolve the problem of fluctuating production of landfill leachate. Moreover, scaling up and down can enrich this kind of reasoning. Recirculation is an *inter-scalar vehicle* (Hecht 2018) that crosses scales: from the micro-scale of a landfill bacteria to the macro-scale of the water cycle.

There are two coexisting translations of the CE concept in Czech public discourse. One is a direct translation—*cirkulární ekonomika*. The other is *oběhové hospodářství*, which attempts to label CE with local terms. It uses the term *hospodárný* (hospitable), meaning being generous to guests and managing resources well, to refer to the economy (cf. Hájek, Kaderka, and Nekvapil 2019: 43). *Oběhové*, in turn, is an adjective form of *obíhání* (to run around), which is frequently imagined as a movement of planets around the Sun or a movement of hands around the clock. Nonetheless, one of the meanings of *obíhání* is evasion, as in running around while avoiding interfering with something in the middle or representing an obstacle to the movement. The second translation underlies what I argue to be the principle ambiguity of circular economy, which is its explicit aspiration to solve the problem of unsustainable economies and its implicit tendency to slide into avoidance of addressing the primary causes of the crisis, which CE is supposed to prevent: namely *market fundamentalism* (see Block and Somers 2014).

A critical scholarship already suggests that CE should be treated with caution. Despite CE's emphasis on waste prevention and resource conservation, several issues deserve attention. Although CE is a heterogeneous field of ideas, its promotion within the EU has been associated with the neo-classical economic theory when nature always waits out there to be measured and evaluated (Gregson et al. 2015: 220). Such a valuation is predominantly conceptualized via quantifiable exchange value with the usual economic heroes of cost, benefit, supply, and demand that lurk around. Since the dominant CE visions are stuck in market fundamentalism, they have difficulties in accounting for the domestic economic activities (Jehlička et al. 2020; Vávra, Daněk, and Jehlička 2018), materials stockpiled (Moreau et al. 2017), informal activities of the actors such as waste pickers who recover materials without being acknowledged (Millar 2018; O'Hare 2018; Zapata Campos 2013). CE may *cement* social inequalities (Carenzo and Becerra 2024) and disrupt existing reuse and recovery regimes (O'Hare 2021). All these dimensions can be boiled down to the point recently formulated by O'Hare and Rams, who say that "framings of the circular economy have thus far remained oblivious to wider social concerns" (2024b: 9). As Corvellec et al. (2022) demonstrate in their review, CE is less radical than it claims to be. It puts into its center a neoliberal regime of governance at the expense of the social dimension and realistic estimations of environmental consequences (Corvellec et al. 2022: 429). In the following section, I will examine what the actual circulation does and what consequences it generates over decades of practice.

RECIRCULATION OF WASTEWATER

The management of landfill leachate was an everyday task at the Pureland landfill. It was not a pleasant activity. Two people had to pull a fire hose lying at the edge of the tipping zone. Since the other end of the hose was coming from the sump situated way below at the foot of the landfill, pulling the hose up was heavy labor, especially during the wet weather when the slope became slippery. Once one of the workers turned on a pump, the other one started spraying the liquid over the landfill's surface. The workers always waited for the manager's order to start spraying and often complained about the manager's insistence that the fire nozzles had to be held in their hands to have reasonable control over the direction of spraying. The workers would breathe in the leachate's droplets when the wind blew unexpectedly, which was not unusual given the position of the tipping zone on the top of a slope unprotected from the wind from the three cardinal directions. I still remember Peter's laugh when the droplets hit my face for the first time, and I tried to wipe the liquid from my face with a sleeve. He laughed while noting that this was normal, and one just needed to get used to it. The workers exposed themselves to the polluting agency of the leachate every day. It is evidence of the uneven distribution of toxicity that affects the bodies of those who take care of garbage on behalf of the rest of us. This unevenness is not a mistake in the system, it is rather its constitutive part that maintains the whole system (Liboiron, Tironi, and Calvillo 2018: 333).

Recirculation is a containment strategy (see Reno 2016b) to ensure the landfill's matter stays within its parameters. It is part of infrastructures for waste management that enable accumulation and prevent overflows and leaks (Stamatopoulou-Robbins 2020: 58). Recirculation has been used at landfills for over half a century (Knox et al. 2018: 691). Technical literature gives different reasons for recirculation. Beyond its importance in preventing the overflow of leachate, it contributes to the stabilization of waste mass, the stimulation of methanogenesis, and the control of dust and fires (Knox et al. 2018). Nonetheless, the frequent use of recirculation worldwide can be explained primarily concerning its convenience and reduction of costs (Knox et al. 2018: 691; Teng et al. 2021: 6). The cost-cutting logic behind recirculation is also evident in research that includes treatment costs as a critical parameter of leachate treatment technologies (Calabro et al. 2018; Hendrych et al. 2019).

To get a better sense of leachate's flows, it is helpful to describe the structure of sanitary landfill. Although sanitary landfills started being constructed in the 1930s in different parts of the world (Melosi 2005: 153), in Czechoslovakia they became common only in the early 1990s as a response to raising awareness about environmental pollution after almost four decades of socialism that paid limited attention to the spread of pollutants into the air and water (Carter 1985; Pavlínek and Pickles 2002). Sanitary landfills in the region usually have layers of concrete, bitumen, and plastic liner at the bottom (Kizlink 2014: 127). In addition, gravel and old tires protect against ruptures of the lining from the inside and act as a bed for the drainage system. These materials are supposed to contain the waste and render it determinate (Hird 2012: 465). Such a struggle for determinacy is supported by a system of pipes that drain it, and sumps for its collection (Kizlink 2014: 127). Leachate must be pumped off from the sumps to prevent its overflow. Once pumped off, it can either be sent to a leachate treatment plant for processing or recirculated within the landfill (Knox et al. 2018; Teng et al. 2021). In the first instance, leachate becomes part of a water cycle because its purified version is discharged into natural waters. In the second instance, recirculation introduces a smaller version of circular movement, but evaporation enables leachate to become part of the water cycle again. Eventually, water originating from the leachate returns to the landfill through rain and water in the garbage, especially biowaste. Beyond these two human-managed processes, limited amounts of leachate evaporate from landfills through the pipes along with landfill gas, which is a mixture that contains methane, CO_2, and other compounds, including water in its gaseous state (Alibardi and Cossu 2018: 231).

Looking from the perspective of the actors on the ground, I was puzzled by the contrast between leachate's critical role in everyday life at landfills and its absence in conversations or broader discourse about landfills in Czechia. As Hetherington (2004) argues, the presence of absence is a vehicle for understanding social relations. It has become clear to me that leachate's silencing is significant. In official representations in Czechia, such as the webpages of local landfills or waste management companies, educational trails, or media presentations, mentions of leachate are rare. Nor was leachate a topic that the manager would bring up in conversation at the Pureland landfill. The situation at other landfills I became more familiar with was similar. None of the four landfill managers whom I got to know ever raised the topic with me. Some of them were willing to discuss certain aspects of informality at their respective landfills, such as waste recovery and repair, smuggling, tolerance for waste pickers, or tricks with waste classification and reporting (see Sosna 2022, 2024; Sosna et al. 2024). However, whenever I raised the topic of leachate, the managers responded briefly, and the discussion was quickly redirected. This diversion or even denial of leachate as a topic represents an example of Alexander and O'Hare's (2023) *technologies of unknowing*. The managers omit leachate to minimize the chances that problems containing it will be discovered and scrutinized. Since leachate results from water, it can emerge in surprising quantities and get through various barriers (see Strang 2014). It is an uncanny matter, always able to create a surprise. Playing down its presence minimizes the need to explain the causes of leaks or give solutions to surprises. On the rare occasions when leachate emerged in conversation, it was *neutralized* by a reference to labor, the laziness of the workers, or the infrastructure used for the recirculation.

Landfill leachate is one of the most unpleasant substances I have experienced. It is dark, volatile, and stinky. Old hoses with signs of wear and punctures may release drizzles and spread droplets in various directions (Figure 3.2). In the Pureland landfill, the hoses were second-hand donations from fire services, and their age set limits on the hoses' ability to contain and channel the leachate stream in the right direction. Occasionally, it was necessary to climb into underground sumps to clean the drainage system or solve technical problems. The odor of the volatile foggy fumes would get deep into workers' noses and inside their hair. The workers would adapt to this fluid and did not share a sense of disgust that would resonate with Julia Kristeva's (1982) notion of abjection (see also Millar 2018: 59). Instead, they were worried about the potential consequences of the intimate relations with leachate for their future. As Petr, one of the workers, expressed: "All our diseases are from this nasty water here, from the droplets we breathe. For what? We spray it; it percolates through the landfill, we pump it off, and spray it again. How much sense does it make to spray when there is no dust?" The workers

Figure 3.2. Punctured fire hose with leachate splashing around during workers' spraying. © Daniel Sosna.

would refer to the potential harm that waste management inevitably entails (see Doherty 2022: 26).

In addition to the unpleasant nature of spraying, the workers were not convinced of its utility. Spraying leachate over the landfill's surface during the hot days made sense because it reduced the dust. However, during the rainy days, when the whole landfill was soaked with rainwater and there were puddles and mud all over the place, it was hard to digest the manager's order to spray again. The workers understood it as a kind of revenge from the manager for their occasional misbehavior rather than a rational decision to prevent leachate spillage and save money for its processing. The manager's insistence on holding the fire nozzles and changing the locations only reinforced the workers' suspicion that it was a vehicle of discipline and solidifying work hierarchy. The workers, however, did not pay enough attention to the recirculation's roles in environmental risk management and economic dimensions.

One day in mid-December 2014, the reasons behind recirculation became clearer. When I came to the landfill in the morning, it was cold and foggy. The temperature was 0°C, and there was 89 percent humidity. Despite the weather and limited potential for evaporation, the manager ordered the workers to spray the leachate. He did not want to discuss the reasons but insisted it "must be done." At that time it seemed counterproductive to me because spraying worsened the muddy surface of the landfill and increased the danger that the incoming garbage trucks would get stuck. A few months later, Jindra, one of the workers, suggested that the manager was afraid of the leaks from the sump at the bottom of the landfill. He did not want to elaborate further and advised me not to stick my nose into the matter. When I probed the manager on the subject, he never revealed anything concerning the leaks. Still, he did mention that leachate should ideally be treated at a plant that specializes in hazardous waste. He admitted that "every now and then" he had to use a cistern truck to transport the leachate, which could not be kept at the landfill anymore, for processing and had to pay the high cost. Clearly, he wanted to avoid or at least minimize recourse to the processing. Without any explicit explanation, recirculation was thus the preferable option for management as it solved the issue of large amounts of leachate and did not require additional expense. It was only necessary to mobilize the workers to do the dirty work and make the wastewater circulate.

The manager's dream of recirculation, however, was not perfect. When the degree of infiltration and the amount of collected leachate in the sumps exceeded the recirculation capacity, leachate would *leak*. Such leakage or the need to send the excess for processing shows that a reference to a circle did not fully represent the movement of wastewater. The case reveals some fundamental limits of circular models. Indeed, the engineers who develop models of leachate recirculation do not imagine them as consisting of closed loops. Unlike the managers who tend to think about recirculation as circular because such a model would sweep complexities of leachate under the carpet, they depict a landfill as an entity that accepts inputs from the outside and releases outputs back (see Alibardi and Cossu 2018: Figure 6.1.2). Leachate tends to be imagined as either circulating within landfills or being purified in special facilities that release it into water. Leachate, however, travels well beyond this model to play its role in circulations and transformations at different scales.

MANAGING TOXICITY

The landfill managers must deal with the irregular excess of leachate and coordinate with special facilities for its processing. However, there is a grey zone of decision-making where the boundary between legitimate and illegitimate decisions becomes blurred (see Frederiksen and Knudsen 2015: 2). In the case of leachate processing, there are usually two options. It can be sent to a sewage treatment plant where leachate is treated after being mixed with other kinds of wastewater or to a plant specializing in hazardous waste. The choice of these two options depends on the degree of leachate toxicity. Whether leachate falls or does not fall into the category of *hazardous waste*, and therefore enters a special treatment regime, depends on a decision by certified specialists who take samples and analyze the concentration of key chemical substances. The Czech norm ČSN 83 8036 prescribes taking the samples quarterly, which means that the specialists announce their arrival at a landfill in advance four times a year. The EU Landfill Directive does not provide specific limits for the discharge of leachate into surface waters and instead delegates this responsibility to individual countries (Stegmann 2018: 502). Given the absence of such limits, there are no EU standards concerning the treatment method either.

In Czech legislation, the leaching limit values are fixed so that the concentrations below the limits enable cheaper treatment in sewage treatment plants.[2] In contrast, above the limits requires more sophisticated and expensive treatment based on biological, chemical, or physico-chemical processes in special facilities (see Teng et al. 2021). All this reasoning depends on the threshold theory of pollution, which assumes that lower concentrations of pollutants are acceptable because of the self-purification capacity of natural waters (Liboiron 2021: 57). This understanding of leachate's capacity to pollute only after crossing certain thresholds is widely shared worldwide (Ma et al. 2022).

Limit values are not always as rigid as they seem but, rather, are open to interpretations, modifications, and negotiations that can be examined ethnographically. This nuances Max Liboiron's (2021) critique of the assumption that pollution can be conceptualized through limit values, arguing instead for a more subtle understanding of pollution that would consider qualities that go beyond the rigid technicist perspective. It would mean overcoming what Bob Kuřík (2021) calls the "engineering of the world." In the case of leachate treatment, limit values are effectively lax. Firstly, the fact that different EU countries set different limit values for leachate pollution weakens their aura as universal. Secondly, the limit values and exact requirements for leachate treatment are hidden using multiple technologies of unknowing (see Alexander and O'Hare 2023). For example, information is not widely available, and it requires energy and time to sieve through complex Czech and EU acts, decrees, and norms to find relevant information.[3] Most important facts are often found in appendices or referred to through web-like structures when one has to follow multiple steps to reach the information, creating a barely penetrable information fog. This may be an unintentional consequence of complex institutions and frictions across the EU-national scalar axis or an intention to restrict access to sensitive information. Given environmentalists' conviction that "landfills are time bombs" (Enviweb 2004) and the bad reputation of environmental NGOs among waste industry actors, there might be reasons why this information is not readily available. Thirdly, both the official documents and scholarly publications from natural and technical sciences, which deal with landfilling and leachate, tend to avoid strict language that would impose a causal logic of "if concentration X, then Y." Differential leachate treatment is rhetorically approached using *recommendations* or *suggestions*, types of treatment are accompanied by words such as "usually," "normally," or "it is possible to." The dry and explicit language of science gives way to softer ways to formulate recommendations for praxis. Finally, the rules inhabit the world primarily as a formal discourse. Even if limit values exist *on paper*, it does not mean they govern life on the ground.

During one of the few occasions when I witnessed leachate transport directly, Slávek took me to the sump. There was already a cistern truck waiting for us. Slávek attached a hose to an outlet valve, turned the pump on, and leachate started flowing into the truck. We had plenty of time during the filling to chat with the driver. The driver described his experience with a previous load to a sewage treatment plant: "Man, I was lucky that nobody took a sample [grin]. Otherwise, I would be screwed. I would have to take it to Alita. That wouldn't be good." He was relieved that everything went well and that he did not need to go to "Alita," the special treatment plant for hazardous waste. Avoiding sampling was a way to treat leachate as non-hazardous waste. This topic was sensitive, and information about it was not customarily verbalized. It was exceptional that the driver raised the issue himself. There were, however, other indirect signals that sampling deserved attention. I recall one of the discussions with the landfill manager about sampling of bottom ashes in the nearby incinerator to evaluate its toxicity: "They take samples but the way they select the pieces ... man, I don't know." The manager doubted the sampling procedure. He was willing to do it only because it was unrelated to his landfill, but it was presented as somebody else's problem. Although these are only indirect suggestions about the informality associated with sampling, it would be naïve to expect a clear-cut answer. One of the critical features of informality is its ambivalence (Ledeneva 2018: 5). Representative sampling is one of the benchmarks of scientific research (Hacking 1990: 6), but it is an arena where informality can easily sneak in. After all, if landfill leachate ends up in a sewage treatment plant, wastewater from multiple sources gets mixed, so responsibility for potential high toxicity gets "dissolved." Moreover, the discharged water from the sewage-processing plant will flow away.

Landfills and waste infrastructures, in general, are places where *anything goes*. Although Feyerabend's (1975) use of this expression relates to scientific methodology and knowledge production, it works well for landfills and waste management in general. These are places noted for their tolerance of informal arrangements and practices (Butt 2020b; Gutberlet et al. 2021; Millar 2018; Nas and Jaffe 2004; O'Hare 2020; Resnick 2021). A recent police investigation in Czechia focusing on reclassifying waste as a "construction material" to avoid taxes and fees suggested immense financial damage to the state and municipalities. In a single case of the landfill near Čáslav, the damage was estimated at €162 million.[4] My ethnographic research at the Kryl landfill, which the same company ran, revealed that it avoided paying the taxes and fees for almost 30 percent of its annual municipal solid waste disposal. Estimated damage to the state and municipality was nearly €2 million during a single year. Another informative case depicting a strategy to bypass the normative treatment of waste is the illegal import of 25,000 tons of plastic waste by a single company over little more than one year.[5] In 2015, the Czech Environmental Inspectorate was puzzled by the disappearance of tens of thousands of tons of hazardous waste via complex rotating transactions among multiple companies.[6] The most recent case of the illegal import of hundreds of tons of mixed waste, including hazardous parts, across the Czech border, and its dumping in places not designed for waste disposal suggests that the courage of free riders is still high.[7] A creative approach to leachate management can hardly be surprising in this context.

Leachate management not only offers opportunities for creative solutions but also represents an environmental challenge. Leachate's unpredictability in terms of quality and quantity makes it difficult to contain using recirculation. It confirms Serres's (2011) claim that emanation and percolation inevitably spread pollution in space regardless of how much we try to contain it. As Elana Resnick argues, such attempts to contain are only aspirational and always open to various disruptions (2024b). Moreover, emanation and percolation of pollution have diverse forms, intensities, and effects, which can hardly be constrained by waste's materiality (Henig 2024: 138). Moreover, understanding and facing pollution should abandon universalistic fantasies and pay attention to the particularities of each case, as Liboiron (2021) suggests. Although specialists on leachate treatment use a much more limited understanding of what these particularities entail, they emphasize the unique conditions of each landfill that need to be taken into account (Ehrig and Stegmann 2018).

The particularistic approach to pollution seems reasonable and fair, but its emphasis on the relations with multiple actors or stakeholders has drawbacks. It may have troubled consequences in assigning responsibility for pollution. The multiplicity of relations along which responsibility dissipates (see Laidlaw 2010) makes it difficult to assign it because of its diffuse nature. The relational conceptualization of pollution would, in fact, enable the polluters to hide behind the diffused network of relations (Gille 2013).

EFFECTS OF CIRCULAR MOVEMENT

CE has conquered public space in Czechia. Media expose society to new governmental policies, technological innovations, and community plans based on the principle of circularity. Engaging with CE is a sign of being "in." Imagining future economic relations without circularity seems almost impossible. Various think tanks and institutes have implemented circularity in their agendas. New institutes and startups have emerged to enable the implementation of the CE. Wherever one looks across the spaces associated with waste management, a circle pops up earlier or later. The spinning effect of all those circles almost brings vertigo. Although it is not entirely like Daniel Knight's (2021) disorientation caused by cycles of crises, it does share the capacity of circular movement to produce a sensation of losing balance. One must stay still to prevent the vertigo from overwhelming a sense of stability.

CE is an opportunity, and the market never misses its chance to roll forward. The market can use the abstract notions of circulation to colonize new economic niches. Stephen Gudeman (2016: 23) argues that market economies tend to develop increasing levels of abstraction. CE fits this trend and has a special quality of its own. Unlike complex economic instruments and models in meta-finance, it takes advantage of the familiarity and seeming simplicity of the circle. It does not mystify the economic relations using blockchain or technologies for transferring information in nanoseconds, although it leaves a possibility for their incorporation in CE open. Although the marriage between waste management and the market is long-lasting (see

Laporte 2000), CE seems to bring new expansion of opportunities, growing from a need to develop new technologies and means of management that would maximize the restorative and regenerative capacities of CE.

Businesses have recognized the enchanting power of the circle. Being successful depends on good marketing, and circularity is a clear and pervasive semiotic vehicle to demonstrate that a company is progressive. There are two major magazines for professionals in waste management in Czechia: *Odpady* (Wastes) and *Odpadové fórum* (Waste Forum). During the last decade, the CE concept presented in these magazines has stimulated immense interest and creativity with the potential to be converted into profit. New technologies and practical solutions are presented as fitting into the growing family of CE advancements. Conferences and educational courses promote the CE concept and facilitate networking among entrepreneurs who want to join the exciting movement. Business must roll over the old-fashioned linear imagination without turning back. It mainly takes advantage of the link to sustainability, which is presented as part of CE. There is a little critical reflection that these two concepts do not necessarily have to go hand in hand (Corvellec et al. 2022: 425).

It is striking how much the promoters of CE in Czechia embraced dominant business classificatory logic and representations. It goes without saying that their organizations mimic the structure and labeling of positions known from the corporate sector. Humans are perceived as consumers, and the industry's attention is captured via titles promoting CE while achieving maximization of production. Overt statements present that the main goals of implementing CE principles are cost-saving and revenue increase. One of the early visual depictions of CE presented on the 2022 version of the website of INCIEN, arguably the leading CE force in Czechia, even used euro coins situated above a circle to convey the key message. The promoters of CE rely automatically on the essential premises of the market, which is understood as an uncontested mechanism bringing improvement and the greater good. The promoters of CE can be rhetorically radical while keeping certain economic principles intact (see Niskanen, Anshelm, and McLare 2020). Interestingly, this market fundamentalism is much broader and also shapes similar "alternative" economic regimes such as those based on gift giving (see Stehlíková 2021). Markets not only colonize the rest of economic life through the processes of cascading (Gudeman 2008, 60; Gudeman 2016: 137), but CE narratives also present the market as synonymous with the economy itself. Any notion that the economy was originally tied with *oikos* and its sound management (Gudeman and Hann 2015: 3) is significantly suppressed, perhaps as an untenable archaism. It is a paradox that contemporary research on the self-provisioning of food suggests that it is close to the logic of the circular flow of value: growing plants and raising animals, consuming self-grown products, and returning the food waste through composting and feed for animals (Daněk and Jehlička 2020; Sosna et al. 2019; Sovová, Jehlička, and Daněk 2021). This kind of circulation, however, is less attractive for the promoters of CE because it is based on old-fashioned ideas and practices tied to *quiet sustainability* (Smith and Jehlička 2013), which is only weakly tied to the market and does not provide similar opportunities for growth and profit-making.

The circular models need to be more realistic and reflexive of their own shortcomings and unintended consequences. Instead of taking a breath via the opportunities brought about by the waste-resource nexus and generating a sense of vertigo via marketing strategies featuring materials, technologies, and processes spinning in circles, the CE movement should think critically about the materiality of waste and its broader socioeconomic context. The notion of a perfect circle diverges from the realities on the ground (Alexander and Reno 2012: 5; Gregson et al. 2015: 224). The case of landfills shows that the workers and managers occasionally have to get their boots on, check the sumps, and solve the imperfections and departures from the ideal circle. When doing so, they expose their bodies to harm contingent upon the need to keep waste matter circulating. When one gets one's feet back on the ground from the vertiginous heights of circular spinning, it becomes clear that waste management produces residues and needs places of the final sink. Raffaello Cossu (2018: 86) argues that loop closure in CE cannot be achieved without paying serious attention to residues and their treatment. Nicky Gregson and her colleagues (2015: 235) argue that achieving CE goals would require a radical transformation of the economic order. The above analysis of the Czech immersion into CE suggests that it carries a lot of assumptions from the more conservative understanding of the economy based on the calculative logic, cost-benefit ratio, and essentiality of money. There is little acknowledgment of the critical voices building upon Karl Polanyi (1944) and his persuasive demonstration of the market's shortcomings. The above analysis proves that waste can be a potent vehicle of critique (Corvellec 2025).

* * *

This chapter started with Romain Garcier's (2012) observation that circular movement may serve as a tool to keep the matter in motion as long as possible to postpone the solution of what to do with the unwanted matter. What landfill managers do when they rely on the recirculation of landfill leachate fits this logic well. Spraying the leachate back onto the surface, letting it percolate, captured in the sumps, and eventually spraying it back again is a vehicle of postponement. Recirculation postpones the moment when the amount of leachate is not manageable anymore and must be taken away for special treatment. This postponement saves the company money, which means that it increases profit. Also, recirculation is a way to minimize the danger of leaks and, therefore, potential fines for environmental pollution spreading directly from landfills. The primary motivation for the engagement with the circular principle resonates well with the one lurking behind the circles of CE: the economic prosperity of those willing to participate. Unfortunately, the circular dream is not perfect, and leachate bites back. It distributes toxic harm unequally. It affects the workers and may pollute the environment via suboptimal solutions for its treatment. Any CE version will likely present the same issue of residuality, leaks, and potential harm. Therefore, it is critical to consider these deviations from the ideal circle.

According to Garcier, the imagery of the circular process itself can obfuscate its material and spatial conditions. In the case of recirculation of landfill leachate, the problem of ultimate waste, which cannot be contained even in landfills, disappears. The same is true for the CE models that try to make waste disappear within the loops. The power of the circle is immense, and policymakers and businesspeople are well aware of that. Behind the bright images of slick technologies, digitization, and growing markets, there is a conservative notion of economies. CE pays much less attention to quieter forms of sustainable life that are not overt and do not necessarily generate business opportunities. With its non-market economy, *oikos* is imagined as a locus of consumers or a source of potential laborers rather than a source of inspiration for alternative economic relations and material intimacies. The hegemony of market imagination in CE suppresses this dimension.

A lesson from the recirculation of leachate demonstrates that a circle is like a double-edged sword. Circular movement can be an efficient tool to solve immediate troubles and generate profit. At the same time, its capacity to postpone while only *circling around* enables it to deflect attention from the primary sources of the systemic problem. There is no doubt that CE's attention toward production and a general sense of frugality concerning resource use are positive improvements. They may decrease overall wastefulness in the human management of resources. However, the circle's enchanting aesthetics and seeming naturalness carry a severe danger of creating a sense that things have already been fixed and we can roll forward as usual, just switching from the linear to the circular logic. Many forms of contemporary life are unsustainable regardless of how sophisticated CE models will be, and it is not healthy to create a false feeling that we have a solution. I seriously doubt that we do. Moreover, harnessing the forces of the market and capital accumulation via engagement with circular opportunities may well deepen existing social inequalities and distribute pollution and harm in ways that will not necessarily fit the notion of justice. When circularity miraculously makes waste disappear, one should be alerted that something familiar is happening again. The only difference is the form mobilized to achieve the disappearance.

NOTES

1. Graeber, to my surprise, does not refer the work of Bloch and Parry (1982) that has become one of the legendary edited volumes in anthropology. This omission is even more perplexing in the context of Graeber's subsequent job at the London School of Economics, where they became colleagues only a year later in 2013.
2. Decree no. 294/2005 Coll. (until 2020) and recently in Decree no. 273/2021 Coll. (since 2021).
3. For example, Czech State Norms (ČSN) concerning waste management are not normally available in libraries nor online. Hard copies can be bought but they are expensive. Access to online versions can be secured only through a special commercial database with temporal restrictions on viewing and printing unless one pays a fortune.
4. Retrieved 20 January 2022 from https://ct24.ceskatelevize.cz/clanek/domaci/reporteri-ct-caslav-chce-po-statu-miliony-tvrdi-ze-o-ne-prisla-kvuli-skladce-kretinskeho-25189.
5. Retrieved 15 October 2020 from https://ct24.ceskatelevize.cz/ekonomika/3204284-firma-z-prestic-dostala-pokutu-za-dovoz-odpadu-z-italie-zaluje-ministerstvo#:~:text=P%C5%AFlmilionovou%20pokutu%20za%20dovoz%20odpad%C5%AF,a%20%C5%BEaluje%20Ministerstvo%20%C5%BEivotn%C3%ADho%20prost%C5%99ed%C3%AD.
6. Retrieved 20 March 2021 from https://inodpady.cz/tisice-tun-nebezpecneho-odpadu-zmizelo/.
7. Retrieved 5 April 2025 from https://ekolist.cz/cz/zpravodajstvi/zpravy/nemecka-firma-podezrela-z-nelegalniho-ulozeni-odpadu-v-cesku-je-v-insolvenci.

CHAPTER 4

LIFE ON THE MARGINS

> 1907 ... Dozens of pairs of blackened women's hands are picking stones from the excavated mass so that only clean coal will reach the wagons and carriages. What is left will be hauled to the heap. The most wretched of the wretched, whom no one would call other than opunčáři, are already lurking there. They swoop on each new batch and, like gold miners, rummage through the piles of slag, from which, with luck, they pull out forgotten bits of coal.
>
> —Karin Lednická, *The Leaning Church* 2020: 223

In early June 2021, we collected scrap metal at the Smoke landfill, a medium-sized sanitary landfill for municipal solid waste, with a group of three waste pickers that I had known for about a year. We carried plastic bags with our finds, going uphill to check the fresh piles of garbage dumped at the top of the hill. One of the piles looked boring because it consisted of biowaste: various weeds, twigs, wilted flowers, and grass. Suddenly, one of the pickers, Libor, started removing grass and discovered a wall of a large metal object. He called the rest of us to help him. After a few minutes of cleaning, we realized it was a large iron container commonly used in construction. The three guys quickly started discussing what to do. They estimated the weight as between 90 and 100 kg, which made this very attractive but challenging to deal with. Tono evaluated the situation: "He [i.e., Ruda, the driver of a front-loader] is gone, so there is a chance. He will not return from lunch during the next half an hour. Let's do it. Fero, take care of that. And you [i.e., me] might help him if you want."

All the guys watched me carefully to see whether I would get scared by the task. At that time, they were unsure how much they could trust me. The employees of the landfill, who collected scrap metal informally as well, would understand our act as a transgression of the silent social contract between them and the waste pickers, who were just tolerated. Also, the pickers wondered whether I would respond to the challenge and demonstrate that I could handle such a large and heavy container. After a quick evaluation of the possible consequences, I decided to participate. We always carried things out of the landfill anyway. The main difference was the unusual size of the object and the fact that Ruda would likely claim this container. Fero and I lifted the edge of the container and overturned it. The fence delimiting the landfill border was about 60 meters downhill. We repeated the lifts and overturns multiple times, and after about ten minutes we had moved the container close to the fence. I could not imagine how Fero planned to get the massive container over the fence. Of course, he managed the task as always. He lifted the lower part of the wire mesh so much that we could push the container underneath. Clearly, he knew well how to deform the fence and had the necessary skill to get things through. We pulled the container further to hide it behind small spruce trees, sat down, and enjoyed accomplishing the task. We left the container hidden in the woods and returned to the landfill to search for more scrap metal. At the end of the day, Fero loaded the container onto a two-wheel cart and pushed it to a scrapyard, where he sold it.[1]

As the story shows, the container was initially supposed to disappear into the hungry mouth of the landfill, which could swallow pretty much anything. We altered the trajectory of disappearance by temporarily hiding the container from the watchful eyes of the landfill employees. This kind of disappearance fits into spatial techniques of waste disappearance proposed by Catherine Alexander and Patrick O'Hare (2023) and adds a temporal dimension. This disappearance via separation and hiding was a tactical tool to reclaim the container while preventing others from taking potential advantage of it. It required hiding it somewhere for a while to keep the claim and extend the container's life as a thing. The owner of the scrapyard found the container too good to be scrapped. Instead, he decided to use it for storage in his yard. Therefore, the container only disappeared temporarily, like Wölfle Hazard's (2022) underflows, to resurface in a new context. Disappearance was only a step leading to the re-emergence and new use of the container.

This vignette takes us to the world of waste pickers who collect primarily scrap metal at the Smoke landfill. They occasionally visit other places for waste dumping as well, but the Smoke landfill is their focal point. As in the previous case of the waste workers, the waste pickers' scavenging is based on the ability to recognize value and rescue it from the morass of garbage. Although the Czech Waste Act describes a need to prevent any *theft* of garbage from landfills, scavenging is rarely understood as such. It is tolerated in the case of the employees and the visitors (sanitation workers, truck drivers, and people who bring discarded materials from their homes and gardens) who dispose of garbage at landfills. Visitors get in via official gates and pay the fees for disposal.[2] Both employees and visitors are part of the expected movement pattern of people and the matter. They take advantage of their presence at landfills with abundant resources and take what they like when an opportunity arises. Employees control the situation and focus on scrap metal and other valuable finds. Visitors usually take things that employees do not want (e.g., already have plenty) or are unaware of.

Waste pickers are in a different position. They come from the outside but not through the gates. Unlike employees and visitors, they have no official permit for being there. At the Smoke landfill, waste pickers take advantage of the holes in the fence at the side of the landfill adjacent to the woods. The trees and absence of people create opportunities for entering the landfill and carrying objects out without being monitored. This does not mean, however, that they scavenge in secret. The landfill manager, front-loader drivers, employees at the gate, foresters, and even local police personnel know very well about waste pickers' activities. It is an open secret in Ledeneva's (2011) sense. Everybody pretends that nothing like that exists. The pickers move across the tipping zone, greet incoming drivers, advise where to tip the loads, and sometimes even help the non-professionals unload their cargo trailers. Some pickers wear work trousers, which they received during the short periods of their official employment, and end up mimicking landfill employees' appearance. Non-professional visitors consider waste pickers to be employees, and professionals take their presence for granted. Waste pickers simply constitute an inherent part of the wastescape.[3]

In this chapter, I argue that salvaging practices preventing garbage from disappearing inside landfill bodies enable the waste pickers to defend their visions of a good life based on the ability to shape social relations tied to work, organize labor, and engage with time in specific ways. It allows them to protect their capacity to act according to their own will in a world full of structural constraints and economic pressures. Picking garbage at the landfill is not a way to escape the hegemony of market relations because it is submerged in them. It is a necessary subsistence activity, which contrasts sharply with the landfill workers I described in Chapter 2. Without salvaging materials and things, the pickers' families would get into trouble and have to find other ways to meet their needs. Nor do the pickers dream much about overthrowing the system, perpetuating social inequality. They have their feet on the ground. Rather than fighting the unjust social order, they search for its cracks and try to use them for their benefit.

Waste pickers and their activities are inherent to waste streams and the larger economy as such. They inhabit patches that Jacob Doherty (2019) calls para-sites for their quality of being beside but intimately tied to the official waste infrastructures. Doherty's main point is that these seemingly strange and heterogeneous beings, practices, and places co-constitute the conditions the larger system relies on to make contemporary life possible (Doherty 2022: 88). While the same holds true for other actors in waste management, such as sanitation and landfill workers (Nagle 2013; Reno 2016a), waste pickers are simultaneously within and outside. Being beside is one way to conceptualize the waste pickers' situatedness. Another way, which is closer to my thinking, is to see them creating and inhabiting the fissures and interstices of the dominant socioeconomic orders (Acedo 2015; Brazzabeni, da Cunha, and Fotta 2015b: 7), seeing them in-between. I propose to unleash a metaphorical imagination of fluid waste streams where waste pickers thrive in alluvial deposits, cracks in the bedrock, or backflows. They are part of the river that carries garbage away but not within its main current. While the main current carries garbage away to make it disappear behind the river's first bend, deposits, backflows, and underflows provide alternative ways to keep garbage present or enable its unexpected re-emergences elsewhere.

What I learned from Czech waste pickers resonates strongly with Kathleen Millar's (2018) notion of a *form of living*, where work represents a central position but cannot serve to reduce human life to its economic dimension. Form of living necessarily includes "a specific mode of inhabiting the world" (Millar 2018: 9; see also De l'Estoile 2014). Although my interlocutors must secure the needs of their families and face various constraints of the racialized Anthropocene (Resnick 2021, 2025), they have choices. Salvaging things from garbage is a form of living that gives them more than just money brought home. It is a way to avoid reducing work into an easily governable domain and workers into puppets directed by somebody else. This is why the pickers tend to come back after short periods of employment, pursuing fortune in MMA fights, or after imprisonment. A sense of freedom and reliance on one's own skills, endurance, and wit outweigh all the discomfort and danger.

While Czech waste pickers at landfills render parts of garbage visible, or sensible in general, they are elusive in various meanings of the word. They rely on the ambiguity of their appearance to pretend that they are employees, sometimes hide in the depressions of the tipping zone or behind the heaps, and other times they run away and hide in the woods. In contrast to other parts of the world where various informal waste pickers, salvagers, and scavengers have vernacular names, there is no established term or even a discourse on the topic in the Czech language. Occasionally, journalists mention that *sběrači odpadu* (waste pickers or waste collectors) did this or that, referring primarily to people who inspect garbage containers in urban areas. Still, activities at large waste infrastructures are silenced. Waste pickers, who have made a living from the garbage at dump sites and landfills for generations do not seem to exist for the outer world. Their not-there-ness stems not only from the relationship to waste that Robin Nagle (2013: 15) noticed among the sanitation workers in New York but also from the informal nature of their work and the small population of these people.[4] They are not only rare but also unrecognized and unacknowledged. While their situation seems reminiscent of endangered organisms tied to specific habitats, the ecological analogy has limits. Waste pickers may vanish from their landfills once they shut down, but it does not mean they will become extinct. The capacity of these people to create new opportunities in the interstices of the dominant economic system is one of their strengths. Therefore, their fading is not a matter of extinction as an ontological issue but a matter of metamorphosis contingent upon their capacity for adaptation.

REDISCOVERING WASTE PICKERS IN THE REGION

The practice of salvaging things from garbage has deep roots in Central Europe. Waste pickers describe that their older relatives, primarily men, have always done that. Expressions such as "my dad did that" or "my grandfather did that" are common. It goes without saying that current waste pickers became familiar with scavenging and tinkering because their relatives socialized them into that environment. Their memories go back to the late socialist times (1970–80s). The history of rag-and-bone men and women, nonetheless, is centuries old. Heike Weber (2022a) describes a collection of discards from households, rubbish bins, and waste piles, a common practice in Central Europe long before municipalities started installing waste management services at the turn of the twentieth century. More importantly, these practices have continued to coexist alongside the official municipal infrastructures (Weber 2022a: 97).

In Czechia, there are historical accounts of the poor at the turn of the twentieth century who were searching for the remains of unburned coal and other scraps at the dump sites in Brno.[5] During the 1930s, waste pickers were documented at the landfills and dump sites in other big cities such as Ostrava and Prague. In the industrial context of Ostrava, waste pickers collected mostly residual coal, wood, bricks, and paper from steel factories, and used the warm slag to sleep at these sites.[6] On the periphery of Prague, waste pickers collected a wide range of things, such as rags, scrap metal, bones, glass bottles, batteries, and toys.[7] At the beginning of the twentieth century, waste-picking men and women could earn a significant amount of money beyond the average salary of that time.[8]

There is another historical line that deserves attention. It is a traditional activity vernacularly called *paběrkování* (gleaning). Before World War II, it was common in the wider region to collect remains of cereals, vegetables, and fruit after the harvest. Michael Stewart (1997: 21, 77) describes vividly the activities of his Hungarian Rom who were collecting unharvested agricultural products, mostly potatoes, in the fields in 1980s. This activity is associated with an internal sense of thrift that drives people to the fields to collect agricultural remains.[9] Although it would be an oversimplification to claim that poverty has been the only driver for *paběrkování,*[10] the unprivileged and poor had good reasons to consider it. These resources were predictable, and food was offered to their families periodically. The history of gleaning is centuries old and goes back, at least, to the Hebrew Bible, which institutionalized the right to agricultural remainders for the marginalized (Bize 2020: 474; Dikovic 2016: 302). A logic of gleaning has played an important role in the socialist times in the Soviet Union, where the households relied on leftover scraps of socialist property (Cherkaev 2023: 81). Although understanding of gleaning may change along with the changes in the postsocialist property regimes (Dikovic 2016), gleaning remains a useful conceptual tool for thinking about relations to resources at different times and in different places. Amiel Bize (2020) argues that the ancient logic of rights to the remainders and their use is relevant for understanding practices that exist in the gaps of the formalized economies *sensu lato,* including the contemporary capitalist ones. Also, Bize shows the potential extent of the economies of remainders going way beyond the collection of wheat stalks after harvest. Waste picking has much in common with the original agricultural economies of remainders. One takes advantage of the things that others left behind. When Czech landfills became more affluent with the increasing consumerism and emerging *throwaway spirit* (see Packard 1960) during the 1990s, social welfare weakened with the onset of *wild capitalism* in Central Europe, and garbage sorting was in its infancy, so taking advantage of the leftovers disposed of at landfills made perfect sense. It offered rich sources of scrap metal, discarded devices, clothing, and even packed food.

Tomasz Rakowski's (2016) ethnography of the Polish poor situates scavenging and scrap metal collection in the broader field of *hunting and gathering* emerging as a response to economic hardship. These strategies included informal disassembling and scavenging of abandoned industrial infrastructures, coal mining, berries, mushrooms, herbs, firewood collection, and poaching. Rakowski demonstrates the creative capacities of people to engage in these quiet and invisible practices to improve their income. In Rakowski's view, this wide range of informal practices has a causal link to unemployment and poverty. While poverty encourages people to search for opportunities within the interstices of the formal economic systems and to recognize new ecological niches, my research shows a less deterministic relationship. It resonates well with what Millar (2018) discovered in Brazil's very different geographic and cultural context. She argues that *catadores* engage in waste picking not necessarily because poverty does not offer them other choices but because reclaiming discards is a *form of living* that goes beyond economic necessity. *Catadores* tend to stick with or even return to their dump site despite having other work opportunities.

While it is not a coincidence that waste picking becomes more visible in times of crisis, it would be a shortcut to directly link poverty and waste picking.[11] A more balanced view would instead envision only intensifying the practices related to waste that always lurked in the background. Historical accounts featuring Czech waste pickers during the Great Depression in the 1930s and references to waste pickers at the beginning of my research at landfills in 2012 as a potential consequence of the 2008 economic crisis suggest that these periods encouraged waste picking. Nonetheless, my fieldwork among waste pickers happened during a very low unemployment rate (2.9–3.4 percent) in the region. Central and West Bohemia's hunger for labor has been strong during the last decade and attracted thousands of migrant workers from other parts of Europe and beyond (Bernard and Vašát 2015; Sosna and Brunclíková 2019). Finding a formal job was possible because many human resource companies were constantly searching for workers. The problem was almost the opposite: resisting formal employment that would alter waste pickers' lives in ways they were unwilling to accept. It does not mean that waste pickers would not face economic hardship, stigmatization, or uncertainty related to housing. They were undoubtedly under various pressures but preferred a different solution to formal employment. Collecting scrap metal at the Smoke landfill was physically demanding, dangerous, and required resistance to weather conditions, dirt, and occasional stigmatization. Still, it gave the pickers a lot of freedom, flexibility, the ability to select collaborators, and, most importantly, solid income despite a certain degree of unpredictability.[12] For them, the advantages of this kind of work outweighed its disadvantages.

Among the waste pickers I worked with at the Smoke landfill, Roma represented a significant portion (60 percent). Although this setting can hardly be generalized to the whole country, frequent derogative comments about *cigoši/cigáni* (Gypsies), *černé huby* (black mouths), or *bubáci* (bogeymen), which I noticed among the employees at other landfills, suggest that the presence of Roma among Czech waste pickers was common. The negative perception of people with darker skin not only reflects the stereotypical perception of Roma in Central Europe (Ort 2022)[13] but also their position as competitors in the search for resources at landfills. Roma have traditionally managed to create and fill the cracks and interstices in the existing economic systems, engaging in what Micol Brazzabeni et al. (2015a) call the *Gypsy economy*. It is a conceptual tool that stabilizes analytical focus in a quest to contribute to the broader project of understanding the human economy, which goes beyond the interest in Roma themselves. Also, Gypsy economy rests firmly within the wider market economy, similarly to the ways informal practices fill the spaces within the formal system (see Hart 2010: 148). Creating and filling the gaps, however, is not just a natural outcome of Romani life. It is rather a deeply ingrained tendency growing from the histories of the racialized Anthropocene, which has been relying on a tacit link between waste labor and racial formations (Resnick 2021).

Collection and trade with scrap metal, which was the object of primary interest at the Smoke landfill, has a long tradition in studying Romani lives in other parts of Europe and beyond (Gmelch 1986: 313; Okely 1983; Pulay 2015; Resnick 2024a; Saethre 2020; Solimene 2015). It gives Roma the autonomy to organize and perform their work. Resistance to employment for wage, which was a common feature among Roma at the Smoke landfill, was not necessarily related to identity construction (see Piasere 1985) but rather reflected more calculative logic. It was a way to keep flexibility and avoid stigmatization associated with formal employment, which all the pickers referred to when talking about their experience with formalized work.[14] Being at the bottom of the work hierarchy and being a Roma often resulted in stigmatization and bad treatment. Also, it was usual that waste pickers were in debt and faced distraint. Formal employment would result in deductions to the distrainors. Therefore, there were multiple advantages of informal waste picking.

To give the reader a sense of the social milieu waste pickers come from, I will use the example of Fero, who became one of my closest interlocutors. A Romani man, he was forty-six when we first met in 2020. He had seven children and lived in a small bare-bones apartment (*holobyt*)[15] provided by a local town, located ca. 6 km from the landfill. While some other waste pickers used cars to get to the landfill, he used an old bike with a trailer to transport his finds into the scrapyard. Fero's family lived in a single large room, which served as a bedroom, living room, and kitchen together. Other related Roma families lived in the same apartment complex. Housing was a source of the greatest distress for Fero because it was not secure. Local non-Roma people were unhappy with this unruly apartment complex and periodically complained about its existence even though it was located outside the urban zone next to parking garages.

After finishing elementary school, Fero became an apprentice in joinery and managed to finish school. He scavenged with his relatives at landfills whenever there was an opportunity. His life changed significantly when he got into a fight at a dance as a young adult. During the fight, he seriously injured another person. Fero ended up in prison, where he spent twelve years. After his return, he started picking up waste again but also tried various temporary jobs (*brigáda*) as a digger or worker in road construction.

He even worked for a local municipality as a worker who cleaned sewers and helped in a municipal scrapyard. A miserable salary, however, was not sufficient for the large family. Balancing between informal waste picking and occasional formal employment became for him a way to secure at least basic housing for the family while earning extra income via scavenging and seasonal collection of mushrooms in the woods. Fero was frequently broke because of the size of his family and occasional gambling, one of the waste pickers' vices. Because of his ties to the local municipality, which he built during the periods of his formal employment, Fero also knew many people from the waste business and beyond. He was the one who could negotiate with the official employees at the landfill, scrapyards, truck drivers, and municipal police. Also, he was the one who negotiated with the municipality on behalf of the Romani community living in the apartment complex.

Fero was a nice guy, but he had a temper. Raising his voice or facing his opponent was natural for him. He was very proud of his children and often emphasized their wildness and resistance to external domination, especially at school. He called them *indiáni* (Indians) and often emphasized that nobody could mess with them because their reaction could be fierce. As Fero said, "*nenechají si nasrat na hlavu*" (they do not let others shit on them). The metaphor of Indians was a way to point at the pride, bravery, and tenaciousness of the Native Americans depicted in the romantic visions of a German nineteenth-century writer Karl May, who was one of the key contributors to the Central European imagination of Native Americans (see Křížová 2025: 113). At the same time, the metaphor carried a more subtle meaning related to racism, injustice, and the uncertain fate of the people who differed from the rest. Fero's pleasure in presenting his children as Indians was a way to defend dignity and independence in a world that marginalized them. Fero was a realist and did not believe they could escape the structural conditions of marginalization. A sense of injustice came from time to time in our discussions during work when Fero complained about a single white mother who received a small flat from the municipality, having only one child, while Fero's family was bypassed. Another strong feeling of injustice emerged during the COVID-19 pandemic when the schoolteachers received a limited number of tablet computers to distribute among children from low-income families but refused to give them to the Romani children. The pandemic made the existing social inequality even more palpable; it *magnified* it (Abu Ghosh 2023: 250). Given the experience of stigmatization and injustice, Fero's identification with the Native Americans was a way to relate to other people's marginalization while building symbolic capital via borrowing their virtues and projecting them into Fero's life.

There was a link between Roma and non-Roma waste pickers. Most of them knew each other from prison. Although not everybody was willing to go into detail about their past, the theme of prison resurfaced repeatedly during my fieldwork. In some cases, this shared experience led to the incorporation of a non-Roma waste picker, Hynek, into a group of Roma. When I asked carefully about Hynek's affiliation with the Romani squad, Fero explained: "He is not *cikán* [a Gypsy]. He is *gadžo* [a non-Roma] from *panelák* [an apartment building]." Later, I learned that Hynek knew the other guys from prison, where they became friends. He was incorporated firmly into the group. He was traveling in the same car, participated in the collection and transport of scrap metal, and also in frequent joking. Hynek fitted in and his membership was understood as natural, even though he was *gadžo*.

However, most of other *gadžos* were perceived as competitors. One of them, nicknamed *žralok* (shark) after a tattoo on his neck, was an enemy of Romani waste pickers. They also knew him from prison, and every time he appeared near the landfill, the Romani waste pickers chased him. The reason was allegedly a bad experience with him as a trickster who sneaked around and stole the finds stored by the other waste pickers in the woods. Žralock's nephew, Milan, who lived in the woods close to the landfill, did not generate the same tensions and was tolerated. Still, he preferred to visit the landfill later in the evening after other waste pickers had left. While some authors try to avoid the concept of scavenging for its potential pejorative meaning (Doherty 2022), it fits surprisingly well with the logic of relations and practices at the Smoke landfill. The dominant scavengers always came first, controlled the resources, or chased those who tried to challenge them or even steal something from them. Once these "lions" satisfied their needs, others could come to see how much was left.[16] It was precisely Milan's position within the web of social relations: being a collector of the "remainders of the remainders."

Current literature on human relationships with waste in Europe emphasizes the role of race (Dunajeva and Kostka 2021; Resnick 2021, 2024a, 2024b, 2025; Škobla and Filčák 2023). Multiple studies convincingly demonstrate an omnipresent and multifaceted link between waste and especially Roma, which Elana Resnick (2025: 34) succinctly calls "waste-race nexus." While I agree with the gist of the argument connecting waste with race, I would like to nuance it because the prominence of the link may depend on epistemic tuning. Asking what the lives of Roma are like differs from asking who does the waste labor. When one focuses on Romani lives, an emergence of race, ethnicity, or another similar analytical category of otherness is a logical outcome of such an approach. The research interests of our team originated from a different point, and therefore have never developed the same degree of attention to or urgency regarding race. Race simply did not grow as a strong theme from our ethnographic experience. Although the Roma are over-represented in Czech waste management, given their proportion in the population, and there is no doubt about their marginalization and stigmatization, race is only one of multiple factors that shape waste labor. In my research, I worked with Romani waste pickers and those who cleaned furnaces in an incinerator. Additionally, I frequently met Roma among the sanitation workers who collected garbage and transported it to landfills or incinerators in garbage trucks. There are, however, many non-Roma people who work in waste management. Landfill and incinerator workers are predominantly white men. Workers who sort garbage in sorting plants or sort biodegradable waste in composting plants are often migrant workers, both men and women. Barbora Stehlíková (pers. com. May 2025) has met only rare cases of the Roma during her ethnographic research in e-waste processing plants and collective systems. It was rather disability that was a prominent factor behind people's presence in these kinds of jobs (Stehlíková 2025).[17] Pavel Mašek's (2022, 2023) research on car breaking does not reveal any affinity to the Roma or racial classification. Roma appeared rarely during his ethnographic research among the car breakers (pers. com. May 2025).

Based on the collective ethnographic experience of three researchers who conducted their fieldwork in different parts of waste streams in Czechia (car breaking, e-waste recycling, municipal solid waste management), the race-waste nexus appears to be less prominent. It would be more appropriate to use the metaphor of magnetism that I explain in Chapter 5. Waste tends to attract other entities with the capacity to disturb, which are preferably kept aside and pushed to marginal spaces. It can be Roma, other stigmatized racial categories of humans, disabled and homeless people, migrants (see Dezhamkhooy 2024), or former criminals who work with waste. Race is only part of the story because other categories of vulnerability are also present. Although this argument does not by any means undermine the role of race in mediating human relations with waste, it proposes recognizing a wider field of relations within which the waste-race nexus operates and accounting for the possibility that the racialization of waste labor may be prominent to varying degrees in different parts of Europe.

* * *

Waste picking at the Smoke landfill was a reasonably profitable activity. Each waste picker normally collected between 3 and 10 kg of pure copper, smaller amounts of brass, stainless steel, and variable amounts of iron. Roughly, scrap metal collection generated between CZK 500 and 1,000 (€20–40) per day for a picker, plus there were other finds that could be used at home, given to friends, or even sold. Sometimes, I witnessed exceptional days of abundance. I recall a day when we, a team of two waste pickers, found 50 kg of copper cables, four brass components from water taps, three air-conditioning units, and several aluminum ledges. Careful search in discarded personal belongings, especially clothes pockets, occasionally resulted in discovering forgotten golden teeth, rings, watches, or other kinds of jewelry worth hundreds of Czech crowns. It is important to note that plastic was of no interest, which contrasts with several case studies from other parts of the world where plastic is an important material for waste pickers (Butt 2023; Doherty 2022; Millar 2018; O'Hare 2022a). The garbage was so rich with other materials and things that it did not make sense to collect plastic. When I told my interlocutors about plastic collection in other parts of the world, it raised surprise. It sounded unbelievable to them that somebody would collect such marginal material, which is so light and valueless. Some pickers doubted that I was serious despite my assurances that I was not joking. Plastic collection at landfills seemed too fantastical to them to be true.

The past of waste picking was perceived as superior to the recent time. One of the Romani waste pickers, Libor, claimed that his father-in-law could earn double or triple the average salary per month by collecting scrap metal twenty years ago because some of the companies did not care much about recycling and occasionally dumped large amounts of scrap iron or even copper at small dumps at their premises. The companies in a nearby town have been producing various engineering products, electronics, bikes, and, later, automotive parts, so the industrial infrastructure created affordances for waste pickers in the region. If one could access these dumps, one could find valuable finds there. While one must take the quantitative estimates concerning past activities of waste pickers with caution, there are two reasons for not wholly rejecting the claim. First, Libor was the least boastful person among the pickers with whom I collected garbage at the Smoke landfill. Over three years, when we interacted, he lost the motivation to impress me with the stories about the heroic deeds of their group and its spectacular finds. He tried to keep his feet firmly on the ground and focused on the collection (Figure 4.1). Second, the stories of landfill workers (Chapter 2) resonate well with this pattern of deterioration of the richness vested in garbage over time.

The predominance of men among waste pickers at landfills was striking. It was accompanied by a sense of hypermasculinity similar to the one that David Kideckel (2008: 172) captured in the narratives about the work of Romanian miners during state socialism. Czech waste pickers put a strong emphasis on their physicality: strength, endurance, resistance to weather conditions and dirt, ability to intimidate others, and masculine sexual potency. Without exaggeration, it is a testosterone world. When I asked waste pickers about the absence of women, their almost unanimous response was that the environment full of decaying matter, unpleasant

Figure 4.1. Libor during scavenging. © Daniel Sosna.

odors, sharp objects, and landfill leachate was nothing that the women they knew would be willing to accept. This argument is biased, of course. There are women who sort garbage in sorting plants or e-waste processing sites (see Stehlíková 2024), which is not necessarily clean and pleasant work. Some women work at the gates of landfills and incinerators, registering trucks and doing paperwork, and some work as crane operators at incinerators. However, seeing women among informal waste pickers at Czech landfills is extremely rare and contrasts with the gender compositions in other parts of the world (Gutberlet et al. 2021; Millar 2018; O'Hare 2022). This is why the following text puts so much emphasis on the activities of men. Salvaging things from landfills is simply a world with a strong gender disbalance, although historical accounts suggest that it has not always been the case.[18]

Czech waste pickers represent *unauthorized* people who are, to various degrees, tolerated at landfills. Similar arrangements are known in other parts of the world. In Romania, waste pickers are tolerated because, among other things, the waste disposal companies benefit from their work (Dohotaru 2013: 209). In Central and South America, access to these sites is often possible for waste pickers (Gutberlet et al. 2021; Millar 2018; O'Hare 2022; Zapata and Zapata Campos 2015), although there have been various attempts to regulate it. Millar (2018: 138) describes the introduction of a work vest as a vehicle of regulation that should formalize the work of waste pickers. Similarly, O'Hare (2020: 58) analyzes the formalization process via the organization of labor into cooperatives and associated tensions that resulted in temporary closures to access waste (see also Carenzo and Becerra 2024). Research in Africa depicts landfills and dumps as accessible for salvagers (Doherty 2022: 106) but also demonstrates that waste infrastructures can become vehicles for negotiating power relations (Fredericks 2018). In various parts of Asia, access to waste infrastructures, including dumps and landfills, is intimately tied to social hierarchies (Butt 2023; Dezhamkhooy 2024; Gidwani 2015). Even in China, which emphasizes high-tech solutions in waste management, landfills offer opportunities for the informal collection of garbage (Zhang 2024: 41). In this sense, Czech waste pickers join the global ranks of various collectors, salvagers, and scavengers who take advantage of the things left behind by others and disrupt the trajectories that lead these things toward disappearance. To do it, they rely on a web of relations that exhibit flexibility, reflecting the dynamic of everyday engagements with garbage. Importantly, their activities contribute to a wider market economy of the state and beyond. As Petr Vašát (2023) argues based on his research among the homeless people in Czechia, these people do not represent a passive body of potential workers but a real workforce that contributes to everyday workings of the whole system. This reading resonates with the forms of living of waste pickers I described. They might be elusive, but their salvaging activities recover tons of mostly metals that can be recycled and channeled back to production.

FLUID LIVES

Petr squats on a small platform close to the top of the landfill. He narrows his eyes and occasionally wipes beads of sweat from his forehead. His bald head is turning red, and his wet T-shirt suggests a previously demanding walk through the soft terrain of the tipping zone. No wonder, it is mid-June, one of those days when the noon heat sucks energy from the human body. Because it is a lunch break, all the usual work buzz is gone, and the only sound we hear is the gurgling of the leachate, which flows from a hose nearby and makes a sound when the stream hits the landfill's surface. Suddenly, Ruda, the front-loader driver, arrives from his lunch break, parks his battered Škoda car, and rushes into his front-loader. Immediately after starting the engine, he honks at us, a signal that everybody understands. We are supposed to get out quickly. He uses this warning whenever his company director, state officers, police, or someone unusual comes to the tipping zone.

Petr starts moving and encourages me to increase the speed. When we hear Ruda's work with the gas pedal and see the swift movement of the front-loader, it is clear that he is angry. We start running to the foot of the hill. When we reach the fence, we see Libor and Jožo still up and walking intentionally slowly away to demonstrate to Ruda that they are not going to "dance as he whistles." Their slowness drives Ruda crazy. He opens the window and shouts, "Get out of here Gypsies!" Then, he turns toward the two walking men and moves off forward, with the engine rumbling like thunder. He increases the speed, and the walking men have to run in the soft matter. Their legs get stuck a few times, which makes their runaway less efficient. Ruda does not care, he keeps the speed and increases the pressure by honking. When it comes to the point when we all expect the worst, Ruda suddenly steps on the brakes, keeps sliding a bit, and stops two meters from the runners. While the men feel relief for not being run over, Ruda looks triumphantly at those two little bugs from the heights of his massive vehicle. After a few seconds of staring, he turns right, hits the gas pedal, and moves away. We all gather in the woods outside the landfill and reflect on what just happened.

I am not the only one feeling uneasy. Petr has a pregnant wife at home, and this fresh experience reminds him that bad things can happen at the landfill. To my surprise, the two main actors of the story, Libor and Jožo, look only moderately concerned and definitely not scared. Jožo only comments on Ruda's aggressive behavior with a phrase I have often heard: "His old lady did not give it to him yesterday again." While this ice-breaking phrase helps all of us to alleviate the tension caused by this dangerous event, I do not expect that we can return to the landfill today. I am wrong. An hour later, after Ruda cools down, we rummage through a load of mixed waste from a nearby electrotechnical company in search of copper wires.

This story demonstrates how quickly the relations at the Smoke landfill can change. After three years of interaction with these actors, one of the most striking impressions is a readiness to react to changing conditions. Constant monitoring of the flows of materials, vehicles, presence or absence of people, smoke from burned materials, and people's moods led to changing foci of scavenging, labor, and modification of social relations. Millar (2018) approached a similar lack of rigidity using the concept of plasticity. Her main point in thinking about plasticity is critically examining the contrast between the formal and informal. Millar's critique tackles the notion of economy and directs attention to the dynamic of living with various forms. While her critical interrogation of informality as an absence of form offers a stimulating reflection, I would like to emphasize a different aspect of her thinking with plasticity. She links the qualities of the material collected by *catadores* and the qualities of social relations. A plastic bottle wall can be squeezed and later partially go back to its shape, similar to the relations between people. Plastic as material becomes a metaphor for life captured in her ethnography.

My field experience resonates with this view of the relationship between garbage and the lives surrounding it. Masses of diverse wastes flowing into the landfill afford certain kinds of engagements, practices, and relations. Different waste streams have "distinct material capacities, which shape how these by-products can be managed and the uses to which they are put" (Reno 2015: 561). At the same time, one should not overestimate the role of materiality itself because it may preclude one from recognizing kinds of relations and scalar registers not directly related to the materiality of waste (Butt 2023: 110). Instead of examining only the qualities and properties of the discarded materials, I pay attention to their *flows*. I join Veronica Strang's (2014) (see also Krause and Strang 2016) call to use water as a tool for thinking about human-nonhuman engagements. Water's fluidity and connectivity, as well as its transformative, vanishing, and emerging capacities, lend it a unique potential for imagining unfamiliar processes and relations through the familiar dynamic of water. Although material qualities and properties certainly shape experience and emotions emerging through bodily engagement with discards (Butt 2020a; Sanchez 2020), the lives associated with the landfill are contingent more upon changing *composition* and *strength* of waste streams rather than individual objects. What drives these lives are the kinds of materials and things that come, their proportions within a waste stream, and the total amount of garbage that gives the stream its strength. Thinking with flows enables us to imagine valuable copper "dissolved" in a load of paper cuts, waste pickers "swimming" in a muddy "wastecourse" of social relations, or material "sediments" of scavenging left in the woods after the activity stopped; all this offers an opportunity to rethink and reimagine entanglements of garbage and humans. The case of the Smoke landfill offers ethnographic material for such rethinking and reimagining.

Generally, the employees tolerated waste pickers. There was a notion of acceptance that they were around and scavenged, and one should not have prevented them from salvaging things from garbage. It would not only add an obstacle to their already problematic lives but also be almost impossible to stop them from coming. Instead of calling the police, the employees, especially the two front-loader drivers who operated at the tipping zone, decided to tolerate them. The silent deal was that the pickers were supposed to disappear immediately when the drivers told them to do so. This request was motivated by occasional inspections of people unfamiliar with real life at the landfill and could have less appreciation for informal arrangements. Also, the pickers were not welcomed immediately after the fires. There was always suspicion that they caused the fire.[19]

The general sense of tolerance, nonetheless, had its limits. I recall a strange reaction from one of the front-loader drivers once I came to the tipping zone with Fero in the morning, and the driver was unhappy to see us. It was strange because we scavenged with Fero two days before and during three days of the previous week, and nobody showed any sign of dissatisfaction. It was more than six months into my field research, so it was unlikely to result from my intrusive presence. Fero explained to me: "There can't be too many of us." I realized that four other waste pickers were already walking in the tipping zone or lurking at the top of the hill. I learned that the employees felt there should be a *threshold* for their tolerance. The tolerance of scavenging was not a question of principal acceptance or rejection but rather a result of negotiation about the degree to which it can be accepted. The employees became nervous when the number of waste pickers rose to over five. Two or three waste pickers could be easily confused with the landfill employees, but when the number grew, it generated concerns. Informal workers became suspicious elements of the dumping landscape.

The second component of the deal was that the pickers could collect anything but iron. Ruda specialized in the collection of iron. Therefore, waste pickers could not compete with him. For Ruda, the iron collection made perfect sense because of its relationship between materiality, abundance, and value. Iron is slightly lighter than copper but is much more abundant among the discards and natural ores. Therefore, iron's exchange value is much lower than the value of copper.[20] Iron collection makes sense financially only if one collects large amounts of it. Also, iron is inflexible, and large objects cannot be transported easily without cutting them into smaller pieces or having large vehicles for transport. Ruda's formal employment gave him the advantage of using the front-loader to transport large pieces from the tipping zone, and he could also use his company's van to transport the scrap iron further.

In addition, Ruda could identify the large pieces of iron from his seat in the front-loader. He did not have to walk through the tipping zone and dig into piles to find small pieces of copper, aluminum, or brass.

Waste pickers were in a different situation. They had to move everything a few hundred meters in bags or with their hands. They had to manage the transport over or below the fence. Moving the material at the end of the day was one of the most physically demanding activities. The pickers had to carry the plastic bags full of scrap by hand. Sometimes, they could use Fero's cart in the woods, but it was still demanding because of roots and stumps. Moreover, Fero and Petr used bikes and carts to transport their finds three kilometers to the scrapyard (Figure 4.2). Transporting large quantities of scrap and bulky pieces was physically and logistically challenging and potentially visible. If one smuggles scrap metal from a landfill, one does not want to attract much attention. Smaller quantities of valuable and flexible copper were much more attractive than the large, bulky iron pieces. One could usually make a few "balls" of copper wires, hide them in a bag, backpack, trunk, or cart, and deliver them quickly to the scrapyard without generating much attention. Ruda could do the same easily, even for the large pieces of iron, because of his position in the system. He could transport hundreds of kilograms of iron from the landfill every month without generating any attention.

Because of these different capacities, Ruda focused primarily on iron and let waste pickers collect everything else. In exchange for his tolerance, Ruda expected help, which included notices about large pieces of iron hidden in the garbage, subsequent freeing of these pieces from the comingled assemblages, and on-site transport. The pickers walking through the tipping zone and rummaging through the piles of garbage were more likely to find iron. They had no trouble helping Ruda with the large iron pieces, and Ruda took advantage of this. The pickers often had to lift and carry heavy pieces of iron for Ruda.

The ideal model of coexistence, however, was fragile. It was immersed in the fluidity of everyday ethics comparable to the one I described in Chapter 2. Waste pickers always waited for an opportunity to outsmart Ruda. Every time he left for lunch, or when he was not at work at all, some pickers started collecting iron. They did not entirely follow Ruda's expectations to throw the fresh scrap iron on Ruda's pile during his absence. Every now and then, a piece of scrap iron could be added to Ruda's pile to pretend compliance with the silent social contract, but most of the finds were quickly smuggled into the woods and out of Ruda's reach. As a reaction, Ruda always tried to store as much iron as possible inside his car's trunk or weighted down his pile with a bucket of the front-loader to prevent any "theft" (Figure 4.3). This vernacular notion of theft differed from the legal

Figure 4.2. The cart with scrap metal ready for transport. © Daniel Sosna.

notion, which prevented scrap iron transport from the landfill. Both Ruda and the pickers were smuggling material out of the landfill. Ruda only used his formal position in the system to legitimize his right to scrap iron and perceived the violations of this claim as theft.

Figure 4.3. "Theft" prevention. Scrap metal compressed with the bucket. © Daniel Sosna.

Ruda did not respect the social contract entirely, either. When he found a large amount of copper cables, which were the primary target of waste pickers' interest, he collected them. It was at times when the pickers were not around or close enough. Ruda would immediately hide the cables in his trunk and sell them in the scrapyard as "unpeeled cables" for a lower price. Alternatively, Ruda negotiated with the pickers to sell the cables to them for a lower price and let them do the dirty work of burning and peeling. Needless to say, the pickers were furious about Ruda's contract violation. Ruda, nonetheless, was not happy either. Stealing his scrap iron, smoking, and the ostentatious presence of some pickers in visible locations also made him furious.

In the unpredictable world of waste streams, where one never knows what will be disposed of at the landfill, the actors must be ready to react to fluctuating conditions. Material agency of the garbage flows cannot be reduced to pollution but has much wider potential (Gille 2007: 26). As Myra Hird says, "Waste is anything but static and submissive" (2021: 194). While waste can be understood as indeterminate (Alexander and Sanchez 2018; Henig 2019; Hird 2012), the unpredictability of waste streams is at stake in our case. Their unpredictable content and strength allow different ways of floating, swimming, or sinking in these streams. Those who come in direct contact with the flowing matter must constantly adjust their engagement modes. The flowing matter can quickly disappear in the cracks created by employees and waste pickers. Memories of stigmatization or conflicts may dissolve in the everyday flow of matter that never stops.

Every time the front-loader drivers started pushing waste pickers out of the landfill, it caused a countermovement. The Romani waste pickers used two major strategies, which were both based on intimidation. The first was based on verbal insults and ostentatious presentation of masculine strength. In particular, Jožo, the most senior bulky guy, with massive shoulders, a large belly, and sturdy arms, tended to approach the drivers, suggesting that they should not mess with him, otherwise, they would get into trouble.

The second strategy took advantage of the stereotypes concerning Gypsies. Fero described to me his approach as follows: "I know where he lives. He has a small dog, so he has to take him for piss every night. Once I waited for him and told him: 'Leave us alone. Otherwise, I will eat your dog.'" Of course, Fero did not eat dogs but repurposed the stereotypical image of the Roma as dog-eaters to defend his right to scavenge at the Smoke landfill.

Also, the employees knew about the criminal past of most waste pickers, which gave them another good reason not to turn them into enemies. What should be emphasized is the fact that all these mutual negotiations based on threats never ended with any serious physical violence. During my whole research, I never witnessed a real physical fight. Drivers relied on scaring waste pickers with their massive front-loader, verbal insults, and threats to call the director or police. Waste pickers relied on scaring drivers with their bodies and words, taking advantage of imagination related to the potential harm that supposedly "wild Gypsies" can cause. These strategies worked, but their lifespan was limited. Every time there was another attempt to drive the pickers away or when they, in contrast, became too impudent, the power balance had to be renegotiated.

The trickery evidenced in Fero's story deserves special attention. It demonstrates a kind of creativity slightly different from those that enable the Roma to manipulate the system (Gmelch 1986: 325). It is a creative genius that enables people to build upon and manipulate existing stereotypes to maximize their success during interactions with others (Okely 1983: 204). This point is not to doubt that these people face structural inequalities and racial stigmatization that channels them into the kinds of work and forms of living intimately tied to waste (Resnick 2021, 2024, 2025). It demonstrates that they do not succumb to becoming passive victims of the system but use their creativity, wit, and performance to bite back. I observed the tactic of mobilizing the stereotypes when my Romani friends interacted with others or even negotiated with me. They knew about my privileged position as a researcher employed in the capital of Prague and living in his own house in a village.[21] During the early stages of my research, several waste pickers or their wives tried to take advantage of a combination of my friendly approach and assumed richness. They tried to sell me old books, tools, electronic devices, and clothing, or to borrow money from me. In particular, borrowing money was hard to resist. The actors could skillfully elicit images of stigmatized humans facing fines for not paying their debts, being exposed to police harassment, and being punished by the local municipality for being different. Fero was especially a master at depicting the severe consequences of my potential unwillingness to lend him money, even though, for example, the day before we had collected scrap metal worth thousands of Czech crowns together. Using the power of stereotypes and repurposing them to maximize success in social interactions was one of the strengths of most of these actors. Usually, another waste picker approached me and gently warned me not to be too naïve and friendly.

* * *

The fluidity of social relations could be traced among waste pickers themselves. When I started working with them, there seemed to be a clear sense of order and division of labor. The most senior, Jožo, was a clear leader who instructed others. His brother Fero, son Randy, son-in-law Libor, and ex-son-in-law Petr did what he told them to do. *Gadžos,* like Hynek, either followed his orders or avoided the interaction as much as possible. Milan and Shark usually left when they saw Jožo coming or came at times when it was unlikely to meet the Roma squad. Interestingly, both sons-in-law used the polite form of addressing Jožo during their speech. This was quite surprising for me, given the informal nature of economic practices, rough language full of swear words, sexual jokes, and a general sense of freedom. It is very unusual for non-Roma families in Czechia to use such a formal way of addressing senior family members. When Jožo generously invited me to address him informally after more than a year of our interaction, I felt privileged because it was something that not everybody was allowed.

At first sight, a sense of order and hierarchy pervaded the social space. Jožo and his son-in-law Libor collected the most valuable and frequent copper cords and cables. Fero was supposed to focus on less valuable aluminum, rare brass and stainless steel, and iron. The iron collection could happen only in the interstices of Ruda's attention described above. Randy and Petr could collect copper, but they had to pay attention to the preferred directions of movement of Jožo and Libor. Hynek could also collect copper, but he always tried to pay attention to things with use value and not only scraps. He was the most curious person and often spent time sifting through the bags with clothing or electronic devices. Milan and Shark were excluded. At first, I thought that these preferences and associated roles were given and stable. Nothing could be less true than this. As Judith Okely noted aptly in her research among the British Traveler-Gypsies: "permanence is not sought" (1983: 61). It required only time to recognize the omnipresent ability to adapt to new conditions. The following case will demonstrate it.

In June 2021, I arrived at the landfill earlier than others. When Fero arrived and parked his bike in the woods, we started moving to the tipping zone. Fero thoughtfully lifted the fence so I could pass below. On the way to the top of the hill, Fero explained that others probably would not come because they had to repair their car. It was only two of us that day. The driver, Ruda, was in a good mood and worked with his front-loader, which was far from us. As usual with Fero, I started collecting aluminum and iron. When I brought a few pieces to our processing spot, I was perplexed by Fero's reaction: "Dan, throw it away. This is not good. Let's do the papers." Papers was a code word among the pickers referring to good-quality copper cables coming to the landfill inside loads of paper strips from an electrotechnical company. In other words, Fero told me his usual specialization was irrelevant. When I asked him to explain the shift, he was unwilling to elaborate on the topic and only noted: "People collect different things in different days." It took me a while to learn from others that Fero was indebted to his brother Jožo and simply could not do what he wanted in Jožo's presence. Whenever Jožo was at the landfill, Fero had to focus on less desirable materials and follow Jožo's orders to perform work tasks. Fero reacted to the changing conditions at the landfill and modified his tactics accordingly.

Similar shifts in focus were common. Although Libor was a specialist in disassembling electric transformers, which required a lot of skill with breaking porcelain parts and unwinding copper wires from the coils, others did the work when Libor was not around or there were too many transformers available for a single picker. Petr was a generalist and collected almost anything. His primary target was copper cords and wires, but he also reacted flexibly to the changing conditions and opportunities. When Fero was not around, the employees were far away or had already left at the end of their shift, and there were heavy pieces of iron; Petr quickly changed his focus to scrap iron, which also required the consideration of the subsequent logistics of transport.

The fluidity of lives at the landfill reflected the unpredictable influx of valuables, luck in discovering them, and the composition of people present. A general sentiment encouraged fairness in claiming the findings. The Smoke landfill was perceived as an accumulation of stuff that the company enclosed and claimed. Still, nobody, including the employees and even police personnel I talked to, would consider it appropriate to perceive this massive mound of garbage as private property. Not only did the policemen suggest that they did not mind people salvaging materials from garbage, but the large, long-lasting holes in the fences were also material signs of tolerance to accessing the garbage. As waste pickers voiced repeatedly: "Anybody can come here to collect." They perceived the landfill as a commons that can serve as a refuge for the poor and vulnerable people in a sense described by Patrick O'Hare (2022a) in his study of Uruguayan *gateadores*. The whole accumulation of garbage was perceived as scraps open to gleaning. Accessing and extracting waste from the landfill was seen as a right for everyone (see also O'Hare 2022a: 84). Of course, only to those willing to overcome the challenging hygienic conditions. The general sense that anybody could claim garbage was an underlying principle upon which other relations and tactics were built. People could come and search for valuable finds. The finder was respected as a new owner of the find who could decide what to do with it.

This idealized image of garbage belonging to everybody collided with the realities of everyday life along several friction surfaces. The waste company running the landfill had to pretend to abide by the law, which meant forcing the pickers away periodically before the visits from the Environmental Inspectorate. Even the widely shared idea of free access to the commons contrasted with the mechanisms to prevent this access. Employees were not always excited about the presence of the pickers, especially as competitors in scavenging, initiators of fires, and undisciplined litter generators in the woods surrounding the landfill. A request that there "should not be too many waste pickers at the landfill" made sense from the employees' perspective.

However, waste pickers took the idea further and used it to their advantage. Their occasional suggestions concerning the presence of a limited number of scavengers at a time were aimed at controlling access to the resources at the landfill. The logic was evident: the more people come scavenging, the less per capita value can be extracted from garbage. Their overt emphasis on free access to garbage and the simultaneous existence of mechanisms to prevent this access underscore Berliner et al.'s (2016) observation that contradictions are essential features of human life and productive spaces for analyses. The pickers' contradictory acts reveal a clash between a shared conviction that all humans have the right to access waste and a need to secure limited resources vis-à-vis others. It is another example of Gudeman's (2016: 11) claim that the economy is based on a tension between mutuality and self-interest. Although the unit of self-interest was not necessarily an individual but a wider notion of personhood corresponding with a family, it required calculative logic and exclusion. The pickers often negotiated with each other to take a break for a day or more to allow others to collect in the meantime. It was usually after a waste picker was lucky in finding a lot of copper, in gambling, or received payment from a short-term job. Others could ask him to skip the next day.

A repertory of tactics of waste pickers to restrict access to the landfill was wider. They frightened potential newcomers with the dangers lurking at the landfill. During the first few weeks of my work with them, they tried to discourage me from coming. Later, I learned that they used the same tactic to discourage all other potential newcomers. The most pervasive theme was the presence of rats. I still remember the day we walked with Libor over a bank situated at the edge of the landfill, inspecting a ditch below us, when we noticed a rustle and quick movement of the plastic trash. Libor looked at me with wide eyes open and explained that it was a rat. There were plenty of them around, I was told. Then he continued with a story of the oldest waste picker, Albín, who had been bitten last year despite having decades of experience with scavenging. Libor ended his story by describing Albín's funeral and then remained silent. Libor's performance was impressive. He did not know, unfortunately, that I already had years of experience with work at the landfills, and stories like that could not discourage me.

The last aspect of the fluidity of relations at the Smoke landfill relates to constantly strategizing who would do what and with whom. In his study of practices of Romani men in Bucharest, Gergely Pulay (2015) recognized two kinds of economic transactions: *collaboration* and *combination*. The former refers to long-term relations based on mutuality, reciprocity, and friendship. The latter refers to short-term forms of exchange that rely on "negotiation and bargaining, implying a series of risks but also unforeseen possibilities to be realized in such a deal" (Pulay 2015: 137). A constant interplay between these two kinds of interaction provides the basis for life. On one hand, basic assumptions and expectations lower the transaction costs and make life bearable. On the other hand, openness and willingness to experiment leave space for fluid adjustments in the uncertain environment of both Romanian streets and Czech landfills. One only needs the ability to cultivate social relations and nurture sensitivity for the most appropriate reactions given the changing circumstances.

At the Smoke landfill, a seemingly natural unit of labor, an individual occasionally gave way to more collective forms of scavenging. For example, Libor tended to be the hardest-working waste picker, always relying primarily on himself during the scavenging, processing, and transporting of scrap metal. But when Jožo asked him to join forces to collect everything together, he never refused his father-in-law. During these days, they collected scrap metal in their shared bags. A shift from the individual collection to the teams of usually two or three waste pickers happened when one person had luck already in the morning and another one saw a chance to increase his success via the decision to join the lucky person and scavenge collectively. Such a request could be rejected or accepted based on multiple factors. The lucky person had to estimate the potential help of the other person during hauling and burning the cables, whether the other person had a car, and his ability to contribute via collection during the rest of the day. Sometimes, a waste picker invited others to buy his unprocessed scrap copper when he felt he had enough finds and did not want to do the problematic burning. Other times, waste pickers requested the lucky one to show them his bags and sell them the contents. Being a waste picker generally meant constant readiness to react to such changing conditions. At the same time, interactions among the Romani waste pickers showed a general sense of mutuality and willingness to help each other if necessary.

SEMIOTICS OF PRESENCE

The most problematic aspect of salvaging scrap copper is removing the plastic coating from the wires. Copper cords and cables represented the most frequent and most significant category of things collected at the Smoke landfill. To receive a reasonable price for scrap copper, one needed to bring it to the scrapyard as pure as possible. There were two basic approaches to removing the coating from the wires and cables: *peeling* and *burning*.

Peeling requires a sharp knife called a *párák* (ripping knife). Since waste pickers did not bring special tools to the landfill, they relied on the discarded knives found in the garbage.[22] These were usually various kinds of old kitchen knives with longer blades, sometimes broken. Working with *páráks* requires great skill. Peeling was one of the activities that generated my admiration. One had to hold the cable in the left hand and simultaneously cut the plastic coating right above the palm. Needless to say, such an activity was dangerous, especially at the landfill, where everything was slippery because of the omnipresent slimy dirt, rain, snow, or sweat on the hands during warm days. One could easily slip and cut the hand seriously (Figure 4.4). The hands of waste pickers bore signs of past carelessness. I still remember the day when Libor was teaching me how to work with a *párák* and laughed that I was so slow and careful. After cutting, one needs to grasp the two ends of the coating, which was cut, and pull them hard apart, then cut again, pull again, and so forth. It is not only dangerous but also physically demanding. It requires strong fingers and the ability to overcome pain, because the plastic ends cut into the worker's fingers and palms.[23] Waste pickers rarely use gloves, to maintain a feel for the material.

The peeling sites were located inside the landfill and in the surrounding woods. Based on the situation, we either sat at a small platform on the

Figure 4.4. Libor's scar (the line going to the root of the thumb) from an old injury caused by *párák*. © Daniel Sosna.

landfill's slope or moved to one of the processing sites in the woods. These sites were distinguishable by a beaten forest floor and the presence of small plastic or metal fragments resulting from disassembling and peeling of devices brought from the landfill. Sometimes, failed attempts to salvage things resulted in their disposal on the outskirts of these sites. These sites were the primary triggers of my research interest when I saw them for the first time: black plastic strips cut from the electric cords, fragments of porcelain from the transformers, broken radios, sheets of metal, a forgotten bike wheel here, and a sink trap hanging from a tree there. These sites sparked my archaeological imagination, envisioning what such activity sites would look like in thousands of years to come. The material remains of activities served as a vehicle for moving between the past and the future based on the ability to perceive the material presence, which reveal so much about "absent people" and their activities. As Alfredo Gonzáles-Ruibal argued, this kind of sensitivity for material remains, which were left somewhere in the middle of nowhere, opens possibilities to generate "a rough material image of the daily banality" (2008: 250). More importantly, engagement with the material hodgepodge of diverse sharp fragments, dirty surfaces, and abandoned things in the woods can serve as a reminder that there is a raw world behind the neat and orthogonal representations of waste management that permeate public discourses.

When certain kinds of garbage disappear from the landfill itself, they become commodities, gifts, things with use value, or parts of them remain in the liminal zone surrounding the landfill. In a sense, these remains have less value than the garbage itself; they are the refuse from garbage, the waste squared. Archaeologists might approach the remains via the logic of classification, which is supposed to tame the unruly world of waste. Distinguishing different processes, one might elaborate on the already existing categories of primary, secondary (Schiffer 1996: 58), and tertiary refuse (Kuna 2015). In this processual way of reasoning, the remains of scavenging would represent a specific kind of refuse left behind after garbage processing. These remains have their own material properties that can be felt and invite certain kinds of relations (Olsen 2010: 157). Tiny pieces of aluminum foil and plastic may get trampled into the forest floor, get stuck into the soles of waste pickers' sneakers,[24] ashes accumulate in slimy heaps that are pleasant neither for humans nor nonhuman organisms, an eviscerated microwave can become a seat.

Nonetheless, what is more important for my further ideas about disappearance is the capacity of these bits and bobs to take part in the process of signification. Material things have an intimate relationship with signification (Keane 2003), and waste is not an exception. Although Reno's (2014) main point concerning the semiotic approach to waste was a shift beyond the human, his theoretical intervention brought an emphasis on waste as a sign of life in a broader sense. When one approaches the remains of scavenging as signs of life, this opens a stimulating space for understanding the disappearance and reappearance of garbage. Let me present a few examples. The sites, both outside and inside the enclosed area of the landfill, where the pickers process scrap metal, contain knives. Interestingly, there is a redundancy because there are often multiple knives located in different locations at each processing site. One can be stabbed into the soil next to an old wooden chair, another one lies next to discarded plastic water bottles, another one is stabbed into a tree trunk one meter above the ground, and another one with a burned handle is stabbed next to a fire ring (Figure 4.5). The knives are not only readily available for work but also markedly visible. Anybody approaching these sites would notice them. An old ax or hammer can also be part of these visual landscapes. Given the fact that the Romani waste pickers compete with the non-Romani ones and have to negotiate with the landfill employees, it is useful to build respect. At the burning sites outside the landfill, waste pickers temporarily stash their bags in the bushes, hoping nobody will steal their treasures. Also, they must be ready for the unexpected visits of the foresters or police. In all these instances, having multiple knives around and clearly visible to the incomers can be read as a message: do not mess with us! While Reno's (2014) examples build especially upon the indexicality of scat derived from the previous experience with bodily processes, the discarded knives reused as tools build both on indexical and symbolic logics of signification. On the one hand, they signify the capacity to cut. On the other hand, they invite imagination of what can happen if one crosses lines with those who left these knives "on guard." Again, waste pickers, especially Romani ones, turn a stereotype of impulsive and dangerous beings to their own benefit: they scare others to claim space.

Another piece of evidence supporting my argument about material semiotics is the placement of decaying animal body parts at strategic locations. In May 2021, waste pickers brought two huge pieces of the bovine limbs from the landfill and placed them next to an oak on the edge of the woods where they normally loaded their cars. The limbs were cut with a saw, so one could see a pale cross-section of the femora covered with dark flesh, which was decaying. There was something about the massive size and conditions of these body parts that made them disturbing. Interestingly, these limbs remained there for several weeks. When I asked about these strange finds, the pickers just shrugged their shoulders and suggested they knew nothing about these body parts. Nobody wanted to be associated with these polluting things. The oak was surrounded by the remains of previous scavenging, and waste pickers often left their bags with scrap metal or other finds there while they were going back and forth to the landfill or burning sites to bring new loads. The most likely explanation for the presence of the disturbing bovine legs is their capacity to repel unwanted visitors from the place where waste pickers temporarily stored their treasures. The odor of the decaying flesh and its visuality were both disturbing enough to keep away mushroom

Figure 4.5. A knife visibly stabbed into the tree trunk at the processing site. © Daniel Sosna.

pickers or random tourists. These polluting body parts simply increased the potential of waste pickers to claim space via pollution.

Michel Serres (2011) directed attention to a general mechanism of claiming space using pollution. His argument about the similarity between wolves marking their territory with urine and industrial companies appropriating space with their toxic waste is simple but persuasive. Serres's logic of claiming space sounds familiar to waste pickers: "When you leave, it [i.e., spot] will be yours because you will have made it dirty" (2011: 4). Waste pickers' processing and burning sites in the woods use the logic of appropriation with dirt. Fire circles filled with heaps of dark ashes, black trees scorched from the flames and smoke, fragments of metal, porcelain, and plastic bottles thrown to the periphery of these sites or spread along the hauling trails; all this is pollution that evidences a claim to space. A critical part of this claim is semiotic. Potential visitors can see and smell the pollution, which keeps them out. People without direct experience with waste infrastructures are horrified by the dirtiness of these unruly sites and the smell of toxic ashes.[25] Nonhuman organisms, which are generally attracted to the dump sites, do not find these sites attractive either, because of the lack of food. In contrast, the waste pickers like these sites because the trees provide shelter and a barrier for others to observe the pickers' activities directly.

Recently, Max Liboiron (2021) enriched Serres's ideas about appropriation of space via pollution, emphasizing power relations. Colonial claims over land go hand in hand with pollution. The basic logic stems from the ability of the more powerful to pollute and expand into the space inhabited by those who lack power. A tendency to expand in space resonates with the activities of waste management companies that try to expand at the expense of local people. At the same time, waste can become a vehicle for challenging power relations. The Lebanese example of activists sending packages of garbage to corrupt government representatives (Saad and Barnard 2015) demonstrates the opposite strategy: spreading pollution to challenge the powerful. I argue that waste pickers at the Smoke landfill do the same. They spread pollution beyond the premises of the landfill to renegotiate the power relations. I do not know any waste picker who would own his or her apartment or house. Their domestic lives always unfold in marginal places owned by somebody else who usually takes advantage of their precarious lives. Spreading pollution in space—garbage in the vicinity of apartments and scrap metal processing discards near the landfill—seems to have a shared unconscious component: to balance the obvious asymmetry in ownership of those who do not own any place versus those who own a lot. Serres (2011: 42) insists that everyone has a tendency to spread out in space. If we agree that both this capacity to spread out and pollute is universally shared, appropriation does not have to always follow the direction from the powerful to the powerless. The pattern could be reversed. As I described earlier, Romani waste pickers are masters in manipulating existing stereotypes. If others consider them as "dirty Gypsies," it makes perfect sense to use just dirt and pollution to claim space. Spreading pollution in the woods makes visual and olfactory claims over space that would be difficult to claim even by those who formally own the land, the municipality. A local forester confirmed to me during a discussion that they had limited options to force waste pickers out of their woods. The woods in the polluted zone surrounding the landfill have been left without much forestry intervention. The power of natural decay and regeneration is left to embrace and hide the polluting activities of waste pickers. It is "their" space, disturbed only from time to time during the police raids that have little effect in the long run. Waste pickers always come back.

One of the most semiotically prominent signs of waste pickers' activities is smoke. It is not easy to conceal it because the smoke from burned cords and cables tends to be dark and dense. Waste pickers established their sites for burning e-waste in dense parts of the woods, in natural depressions, and often close to sources of water such as creeks or marshes. Although I identified eight sites in the woods, which were used during different periods and by different users, I learned most about burning from a site about 80 meters from the landfill's fence. This site was hard to detect from the outside, because one had to grovel through young spruce trees to get in. The area of several square meters consisted of a central fire ring surrounded by diverse non-flammable e-waste components, burned gloves, wooden sticks, empty water bottles, and pieces of carpets for sitting. Spruce and maple trees closest to the fire ring had their branches darkened, sometimes without pines or leaves. Although this unruly site, which was dominated by a central black heap of ashes growing every week, would likely generate disgust among people unfamiliar with the lives of waste pickers, it was hidden. The trees could even conceal the flames, which are otherwise highly visible in more open environments. Interestingly, there was little concern about environmental harm (see also Pelikán 2010).

At the end of the day, waste pickers would normally bring plastic bags packed primarily with electric cords and cables to this site. A tangle of these components would be sorted, cords with *pozlátko* (internal aluminum foil) would be removed, and the remaining components would be placed on the top of the heap with the ashes. On the top of the tangle one needs to put a piece of polyurethane foam, which serves as kindling for starting the fire. This trick works in all weather conditions, including rain and snow. The foam can be found easily at the landfill. Once the foam ignites the plastic coating of the cords and cables, one needs to use a stick to lift the tangle to secure more oxygen for burning (Figure 4.6). While burning, pickers periodically lift, turn, and beat the tangle to separate copper from other burned plastic. At the end, the copper tangle is removed from the fire, it is ideally cooled with water, and after several minutes of waiting, it is shaped by hand into a copper ball or balls, which are put back into the plastic bag for further transport to the scrapyard.

The greatest challenge is to conduct this process without attracting much attention from the outside world. Large amounts of copper cords and cables generate larger flames and smoke. Also, if a waste picker hurries too much, the lack of patience also results in heavy smoke. To burn the material carefully requires knowledge and skill. One needs to estimate the safe amount to be burned and the intensity of the wind, which affects the visibility of smoke. Also, one needs to know how to manipulate the burning tangle to minimize the smoke. Some waste pickers use soil mixed with leaves to control the intensity of burning. Once they feel there is too much smoke, they reduce the fire by *powdering* it with the mixture. However, one must be careful not to kill the fire entirely. Burning dozens of kilograms of cords and cables daily without attracting attention is a challenge.

A specific dark smoke is an index of burning something inappropriate. Its density and smell signal that this kind of life differs markedly from the disposal at landfills and highly sophisticated burning in incinerators. At the same time, this sign of life reminds us that waste does not disappear as expected.

Figure 4.6. Burning of the electric cords with the copper core. © Daniel Sosna.

In contrast, it challenges attempts to make it disappear via enclosure or containment. It leaks, passes through porous boundaries, and emanates in space (Serres 2011: 44). The smoke challenges the technoscientific aesthetics of enclosure that is supposed to separate humans from their waste (see Zhang 2020: 94). It reveals the lives that unfold beyond the official limits of the waste management systems, while having intimate relationships with these systems.

During unfortunate weather conditions, such as days with low humidity, good visibility, and without wind, waste pickers' smoke can be seen even from nearby towns and villages a few kilometers away. The smoke can almost always be detected from the top of the landfill where the employees work and truck drivers tip their loads. I witnessed multiple times that both front-loader drivers approached waste pickers to advise them not to burn the material in the woods. They were afraid that the smoke would attract the police, who would subsequently scrutinize how the pickers obtained the scrap metal in the first place. Their worries were understandable. The employees could be punished for tolerating the pickers' activities. Still, there was nothing that could stop the pickers from burning. I witnessed one of the police raids during which all the scrap copper was confiscated, and two waste pickers were summoned to the municipal court. I expected that it would be the end of burning. Not at all. The next day, the pickers were back at the landfill and decided to go to the very same site where they had been caught the day before. There was a degree of hesitance about whether it was wise to burn again, but the final decision was to proceed. The logic behind the decision was straightforward. Petr and Libor simply needed money for their families. Processing cords and cables via burning was the most efficient solution to purify scrap copper and obtain money from the scrapyard. The risk of being caught and paying a fine again was less depressing than being unable to take care of their families.

DISAPPEARANCE THROUGH MIXING

One late afternoon in July 2021, Fero's marginal note about burning captured my interest. It was one of those days when one saw the air shimmering above the garbage because of the high temperature and lack of wind. We were quite successful in collecting copper cords, and the question was what would happen next. The woods were dry after a prolonged period of summer heat, so there was a danger of forest fires. Also, the smoke from burning would be highly visible because of the windless conditions. Fero was not eager to burn the cords in the woods. Then he said something surprising: "I'll take the bag home. Tomorrow, I will mix it with grass and burn it together next to our *holobyt* [apartment complex]." When I expressed my doubts about this tactic, he continued:

> I'm gonna visit *měšťáci* [municipal police] first. They are my buddies. I'm gonna tell them that we have the grass cut and there is more stuff from the trees next to the *holobyt* and I am just gonna burn all that. Obviously, there will be a lot of smoke but they will be able to relax because they will know that it is just grass. Everybody knows that grass and leaves make a lot of smoke.

I realized that Fero suggested mixing as a tactic to hide the presence of e-waste and resolve the problem of purifying scrap copper. Mixing was another modality of disappearance based on temporary concealment.

Later, I realized that the activity of mixing had a wider significance. Occasionally, Libor and Jožo came to the landfill and already had a plastic bag with scrap metal. At first, I did not pay much attention to this phenomenon because waste pickers sometimes took their finds home and brought them back to make burning worthwhile. When they had too few cords and cables, they waited to assemble at least 8–10 kg and then did the burning. In practice, it meant coming to the landfill with a bag with content and adding new items into the bag, therefore mixing the older and new items. What was strange, however, was Libor and Jožo's performance. They tried to either avoid the discussion about the content or react in ways that differed from their usual reactions. I recall especially our scavenging in March 2022 when they collected scrap copper together into a single bag, which they had brought half-full in the morning. When they emptied the bag at the end of the day, a few thick cables were inside. While Libor started peeling them with a knife, I wondered about the richness of these finds. These were not common finds at the landfill. When I asked, Libor expressed his surprise that such massive cables "came in a load of construction debris, and nobody had noticed the cables at the construction site." When I carefully opened the topic of the material in bags brought to the landfill to triangulate with other pickers, I always received ambiguous responses about a lack of knowledge of others or suggestions that assembling the material for burning over a period of a few days could be advantageous. Since I was already familiar with how waste pickers communicated, I did not find these explanations convincing. It seemed that there was something that they did not want to reveal, a kind of informality that was sensitive and problematic even for them. I can build here only on indirect suggestions, but it is likely that some of the fresh-looking copper cables came from the functional infrastructures or construction sites in the region. There are two reasons for my suspicion. First, waste pickers are not the only actors interested in scrap copper. Workers in the factories and the construction industry collect it as well, as evidenced by the assemblages of empty cable coatings disposed of at landfills. Waste pickers frequently complained about other people who were at the previous stage of the discard process and took advantage of the scrap copper. Finding such thick and fresh-looking cables in the garbage is possible for the pickers, but it would require a lot of luck. Second, it is common knowledge that the companies and state institutions that build or take care of infrastructures containing copper components face occasional raids and disappearance of these components. Mixing pieces of scrap metal would make perfect sense if one wanted to conceal their diverse origins.

Mixing is a transformative process. It can change the relations among the components of the mixture, their properties, and create new properties of the mixture itself. While the pieces of scrap metal collected into and mixed within bags and later piled, mixed, and burned are relatively homogenous *mixtures* in comparison to *assemblages* proposed by Gilles Deleuze and Félix Guattari (1987), mixing is an active process that operates in both. This process is based on interaction and transformation. The elements of the mixture coexist and affect each other. Different qualities and proportions of the components of the mixture affect its final properties; like the components of concrete (Jamieson 2020). The creativity of mixing is known from a wide range of domains. Cutting drugs, mixing agricultural products of various quality or geographical origins, mixing money collected from various donors, mixing oil, gas, or water transported and stored in reservoirs, an escaping criminal who mingles in the crowd, these are just a handful of examples where mixing is used to achieve various ends. What interests me in particular is the capacity of mixing to conceal and, therefore, serve as a vehicle of opacity.

Waste is a stimulating sphere of thinking about mixing because waste tends to be heterogenous in the first place (Hird 2021: 92) and this heterogeneity has a special relational capacity to refer to diverse places and times (Sosna 2017) despite the goals of waste management to achieve the opposite: keeping waste in the non-relational state of suspension (Reno 2018: 39). The case of mixing discarded scrap metal with metal from non-discard contexts is a way to conceal the origins of the scrap metal brought into a scrapyard. Once copper cables and wires end up in a plastic bag, which is used for the collection of scrap metal at the landfill, it is virtually impossible to distinguish between what was found at the landfill and what came from other places. Moreover, when the bag is turned upside down and emptied at a processing or burning site, the content becomes a pile of material. Subsequent burning creates another step in erasing the link between the origins of copper devices and pure copper transformed by fire. All the surface coatings get burned, and a resulting tangle contains hardly identifiable pieces of pure copper wires and plates. The tangle is shaped into a dense ball, which can be taken to the scrapyard. At that point, a specific register of indeterminacy reaches its peak. It is the register that relates to the conditions of not knowing (Henig 2019: 91; Reno 2018: 38). No one can determine where each piece of copper comes from. Not only does burned copper lack information about its journey prior to becoming part of the tangle brought to the scrapyard, but it is also challenging to untangle a piece from a complex whole in the first place. A piece of scrap becomes an integral part of a whole. This process scales up when the tangle becomes part of larger and larger wholes of scrap copper: at the scrapyard, the train station, and the smelters focusing on non-ferrous metallurgy.

Relations between parts and wholes can be tricky. A perception of what constitutes a part or a whole may depend on the relations viewed from different perspectives (Strathern 1992). Although Strathern's theorizing focuses primarily on persons, bodies, and reproductive processes, it raises an interesting point concerning dismantling. She argues that Melanesian conceptualizations make it possible to dismantle a whole (e.g., a person) and invest the relations, which previously constituted the person, anew (Strathern 1992: 76). It implies that while wholes change over time, relations, which constitute the wholes, last and can be creatively recombined. The relations themselves do not seem to disappear and fall into oblivion. What happens in the world of scrap copper is almost the opposite. The actors try to cut the relations to create the material wholes. Keeping the problematic relations (i.e., the origins of the copper) could threaten the entire salvaging project. Since parts may challenge the notion of a whole (Alexander and Sanchez 2018: 3), turning down their relational capacities can make sense in certain contexts. The physical proximity and materiality are critical in our ethnographic case of scrap copper. Copper wires and other fragments are entangled in a very physical sense: they touch and wind around each other. Their sharpness, presence of both flexible and stiff elements, and entanglements make it challenging to dismantle the whole. Even if there were a sign referring to the origins of the copper prior to burning, which is highly unlikely in the first place, it would be difficult to isolate the fragment and scrutinize it. Although disassembling can be perceived as an activity generating satisfaction (Sanchez 2020), this is not the case for burned copper tangles. Their dismantling would require a lot of patience and time. Since waste laborers calculate whether it is worth investing time in the disassembling and purification of scrap metals (Butt 2023: 85), the physical nature of the copper tangles discourages detailed scrutiny. Moreover, waste pickers make every effort to mix the scraps to make any internal scrutiny difficult. Sometimes, they even fold the tangles in ways to hide problematic elements, such as wires with pieces of unburned plastic or aluminum foils inside.

The case of grass burning shows a similar logic of mixing as a vehicle of opacity. Burning is supposed to conceal the hybrid nature of the pile exposed to the fire. Transformation by fire and entanglement of the copper scraps is supposed to cut any potential relation that could identify the source of the copper like in the previous case. Burning in such a public environment, however, shows yet another dimension of opacity. While most external observers know how to interpret the dark smoke coming from the places in the vicinity of the landfill, "grass burning" is supposed to conceal what is being burned. Mimesis of the traditional practice of burning grass and leaves represents a cover for the polluting activities of the pickers.

* * *

This chapter started with a quotation from Karin Lednická's fascinating writing on the history of mining in Silesia. Her novels vividly depict life in this industrial region, including fragments of stories about the practices of those who were dependent on industrial waste from the mines and steel works. While we know a bit about the hardship of living in the shadows of industrial heaps, a whole gamut of these forms of living remains largely unknown. I tried to contribute to the knowledge of people—often marginalized—who have lived in an intimate relationship with various kinds of waste and leftovers in Central Europe for centuries. Building upon Millar's concept of the *forms of living* I examined why waste pickers from the Smoke landfill do what they do despite having other work opportunities, including formal employment. All the challenges of dangerous, polluting, and physically demanding work associated with stigma, uncertain height of income, and a necessity to transgress official rules are outweighed by freedom, work with real peers, who are often relatives, and the possibility to take advantage of waste pickers' strengths: wit, skill, and resistance.

It is hard to imagine what might stop waste pickers from continuing this form of living. One of the things that amazed me most was the incredible persistence to continue scavenging regardless of what was happening around them. I can imagine only a shutdown of the landfill, imprisonment, or serious illness as the scenarios that could stop them from coming and searching for valuable finds. Nor could my questions about the distant future shake their conviction to continue. I still remember the day when I asked Petr whether a lack of formal employment did not make him worry about his retirement. He looked at me with surprise and responded: "Why? I will not make it to retirement anyway." The way he said it cut me down to size. His utterance was straightforward and neutral, without any sign of self-pity or grievance. It was smooth, without much emotion. It seemed as obvious as a statement that the sky was blue. Later, I realized that every other picker thought about their future similarly. Their imagination had no space for retirement, so it did not make sense to worry about the consequences of their informal work and the absence of pension in old age. It is only a tiny detail, but it brings an insight into their view of life.

Waste pickers shape the disappearance of garbage in multiple ways. Primarily, they disrupt the expected normative trajectories that are supposed to seal garbage inside the bodies of landfills. Collecting, disassembling, and repairing garbage gives new lives or at least extends the life of things. I demonstrated that a repertoire of waste pickers' tactics includes the intentional temporary disappearance of selected garbage items via their spatial separation. This mode of disappearance is reminiscent of an underflow, which vanishes temporarily to appear later somewhere else. I showed that these underflows were contingent upon the changing composition and strength of the waste streams. Fluid relations among the actors reflected the dynamic of unpredictable waste flows. Staying still or insisting on the rigidity of relations would likely fail.

Another modality of disappearance relates to waste pickers themselves. Although each of them salvages a few tons of scrap metal yearly, their existence and contribution to sustainability and circular economy remain unrecognized. These people are virtually nonexistent. To keep their invisibility, waste pickers rely on their ability to mimic the work of landfill workers. Waste pickers tend to wear work apparel reminiscent of the employees and behave in ways that less experienced visitors of the landfill can hardly recognize from those of the employees. Still, waste pickers occasionally need to escape from the landfill or at least hide crouched behind the heaps of garbage or inside the terrain depressions. When the danger of being seen inside the tipping zone vanishes, they reappear and continue their activities.

Waste pickers' activities carry a specific semiotic component. I paid attention primarily to two kinds of signs generated by scavenging: material sediments of scrap metal processing left in the woods and smoke from its burning. I depicted the material remains of scavenging left in the woods as signs of life. All those plastic, metal, and porcelain fragments and dark ashes refer to the salvage activities demonstrating waste's capacity to spread in space. A large part of the former garbage from the landfill disappeared but left a trace behind. The positioning of knives and rotten pieces of meat served as a warning for the outsiders to avoid the territory claimed by waste pickers. Elements of former garbage reappeared to maximize its semiotic impact on those who might disturb waste pickers' activities. Smoke had a unique role in the semiotics of scavenging because it was arguably its most prominent visual and olfactory component. Smoke from burning copper cables and cords revealed lives unfolding beyond the official limits of the waste management systems and brought attention to the people and activities that would be almost invisible otherwise.

The last modality of disappearance is related to mixing. Using the two kinds of mixing, I demonstrated that it represents a powerful tactic of the concealment of things and processes. In the first case, the pickers were mixing electric cords and cables collected at the landfill with those extracted from the infrastructures in non-discard contexts. Waste pickers tried to cut the relations to create the material wholes that could be later transformed via incineration, therefore effectively erasing any sign of their divergent origins. In the second case, mixing electric cords and cables with grass and other plant-based materials during burning, which was meant as landscape management, was supposed to conceal the purification of copper through fire. Both cases show that mixing is a potent vehicle of disappearance when garbage per se does not disappear, but its presence becomes challenging to determine. Mixing weakens the capacity of an observer to distinguish between different components constituting the whole.

NOTES

1. The container had 95 kg of scrap iron, and Fero sold it for CZK 10 per kg. This price was higher than usual for a kilogram of scrap iron at that time (CZK 6 per kg). The owner of the scrapyard valorized the container so well because it could be used in his yard for storage, not just as iron matter that would be sent for recycling.
2. Not everybody enters and pays within the official regime. Garbage can be smuggled into the landfill if one bribes the person at the gate. The rumors at the Smoke landfill

said that an informal fee for smuggling a truck with garbage into it was CZK 500 (ca. €20).

3. While the presence of waste pickers at the Smoke landfill is tolerated, it is not necessarily the case of other Czech landfills. While informality generally pervades waste management in Czechia, some landfill managers are stricter and try to prevent outsiders from getting into their premises. It is usually achieved by a combination of better fencing with barbed-wire, and surveillance systems that monitor the landfill.
4. At the Smoke landfill, there were seven people who were coming on daily basis and about twenty in total. It is difficult to estimate the number of informal waste pickers at all Czech landfills, but the entire population will be in lower hundreds. Therefore, it differs from other parts of the world.
5. *Lidové noviny* 26 May 1909.
6. *Světozor* 1937, vol. 37, p. 830.
7. *Světozor* 1931, vol. 31, pp. 676–77.
8. Gains of mediocre waste pickers per day could reach CZK 60 (*Světozor* 1931, vol. 31, p. 676) in time when the average salary per month was ca. CZK 700. Skilled pickers could earn even more. This suggests that one could earn good money when one was not afraid of the dirt.
9. My own grandfather kept collecting corncobs left in the field after harvest until his old age. He learned this as a child born into a rural family with twelve children. His sense of preventing the loss of value was so deeply ingrained in his personality that he could not stop doing it later in his life despite not being driven by poverty. He had a decent salary and, later, retirement benefits. I have met multiple people like that in the rural environment.
10. Today, there are even activist groups that built upon the heritage of *paběrkování* and try to re-establish it as a sustainable and morally valuable practice.
11. This visibility can be attributed not only to the intensification of the practice but also to the semiotic potential of garbage itself. Berg (2023) shows that discursive emphasis on garbage and its management can serve as a vehicle for deflecting attention from the main sources of crisis.
12. The value of specific temporality and freedom in scavenging became evident when Petr, one of Romani waste pickers, described his experience with working for Amazon. He could earn good money but was horrified by the punishment (financial loss) if a person came back more than two minutes late from a break. This rigidity was simply incompatible with Petr's perception of time.
13. See also Abu Ghosh, Yasar. 2008. "Escaping Gypsyness: Work, Power and Identity in the Marginalization of Roma." Unpublished Ph.D. dissertation, Department of Ethnology, Charles University, Prague.
14. It is an irony that a politically correct label for Gypsies in the Czech public discourse is *nepřizpůsobiví* (inadaptable). What makes them inadaptable in the narrow understanding of adaptation as adherence to formal rules is their refusal to adhere and lose flexibility. If one, however, understands adaptation as a capacity to adjust to changing conditions, it would feature among Romani strengths.
15. *Holobyt* is a vernacular word referring to the lowest possible quality of housing provided by the state. It is a combination of the words *holý* (bare) and *byt* (apartment). The equipment is as basic as possible to prevent any damage caused by inhabitants.

16. In social and cultural anthropology, there is arguably a hesitance to engage with biological and evolutionary thinking unless they are objects of research (for exception see Ingold 2000; Ingold and Palsson 2013). Awareness of various historical slips such as racial theory, phrenology, and sociobiology has resulted in scholarly self-censorship in developing certain kinds of thinking about similarities and differences among the species. The baby was thrown out with the bathwater. Philosophers are less hesitant to examine general principles shared across the species. Michel Serres (2011) demonstrates the power of analysis that positions humans among other species that all appropriate space by pollution. My point of juxtaposing scavenging among waste pickers with the one among the nonhuman scavengers is not to stigmatize certain humans but to point at marked similarities in the relations of scavenging.
17. See also Stehlíková, Barbora. 2024. "E-Waste Between Morality and Ethics: Waste Practices in the Czech Republic." Ph.D. dissertation, Department of Sociology, Institute of Sociological Studies, Faculty of Social Sciences, Charles University, Prague.
18. An explanation of the historical shift from the model where both men and women participated in waste picking at dump sites to the single-gender model is beyond the scope of this book and would require additional archival work. There are, nonetheless, two factors that deserve attention. The first is a hybrid subsistence model of the families of waste pickers. Their families tend to have many children to secure social benefits and housing from the state as a stable source of well-being in addition to the fluctuating income from waste picking. Somebody has to take care of the domestic sphere to legitimize the right to the benefits. This reliance on social benefits was probably much weaker before the 1970s when the new laws concerning social security of families were established. Second, an access to garbage at dump sites has been declining over time because of various security measures. Again, men have been perceived by their families as those who have been better suited for breaking through fences and risking the punishment for their illegal activities.
19. While landfills can catch fire from various sources, including the garbage itself, it is true that waste pickers can also cause them. Most waste pickers I worked with were heavy smokers who had to take frequent breaks for a cigarette and did not always leave the tipping zone where fire danger was high, especially during summer.
20. The contrast between these two most common metals collected at the landfill was striking. While the price for one kilogram of scrap iron fluctuated around CZK 6 during my research, the price of scrap copper was from ca. CZK 100 to 170 based on the quality (burned, thickness, etc.).
21. Owning an apartment or house is hardly imaginable for the waste pickers I worked with. Accommodation was the most critical challenge in their lives. Renting a decent apartment itself was extremely difficult to achieve. Ownership was in a sci-fi sphere. This single fact made me a very different and privileged human being in their perspective.
22. Almost complete reliance on discards, patience in waiting for the arrival of new tools, and creativity in their use represented one of the most fascinating aspects of scavenging and tinkering. Waste pickers did not bring tools to the landfill. Rarely, one of them might bring *štípačky*. All the disassembling, cutting, peeling, unscrewing, hammering, breaking, and bending were performed with the tools and things-to-become tools previously found in the garbage.

23. I must note that this activity was a challenge for me despite having my fingers trained during three decades of rock climbing. Although this note may sound self-centered, the point is to emphasize how difficult it is to separate the plastic from the metal core.
24. Sneakers are the first-choice shoes for most waste pickers despite being unsuitable for the terrain. The seeming "irrationality" in their use is comparable to Jason de León's (2013) observations among undocumented migrants. Sneakers get wet, do not protect from cold, sharp objects, and injuries, and soon fall apart. Among waste pickers, the main advantage is their availability. Sneakers can be found easily in garbage, used for a limited length of time, and replaced whenever necessary.
25. The ashes from the open burning of electronic cables and cords are highly toxic (Xiezhi et al. 2008).

CHAPTER 5

PLACES OF DISAPPEARANCE

> It occurred to me that this street holds nothing but dirt. ... This is a sick artery, where only the toxic elements flow, so its wall gets clogged. It is like cholesterol crystals are clinging to it. So, then more bad stuff gets built on top of this deposit. That is why there are only sex shops, pawn shops, and slot machine arcades around here.
>
> —Pavla Horáková, *A Theory of Strangeness* 2018: 9

In August 2012, I first learned about the plans for the construction of a new incinerator next to the Pureland landfill. At that time, Petr, the front-loader driver at the landfill, was enthusiastic about the plans. He used to work at an incinerator in a hospital for a short period of time, and his experience was positive: "There was only gas evaporating at the end," he said. Unlike many other local people, he was not afraid of the incinerator but saw it as a natural, technological movement forward. Unfortunately, he had not known that the new incinerator would replace his landfill five years later, and he would have to search for a new job. What captured my main attention, nonetheless, was something different. I started exploring why the incinerator was next to the existing landfill. Naturally, there were utilitarian reasons for building the incinerator nearby and taking advantage of the existing infrastructure. The area was fenced, there was a gate with the scale and associated offices, and access to a main road that would facilitate the transport of thousands of tons of cargo. Still, it may probably be better to have the incinerator much closer to the city, especially because of the length of pipeage required to deliver heat to the city. A heating plant, which the company ran as well, was almost in the city center. The existing landfill seemed to serve as a magnet for the new infrastructure that was supposed to make the waste disappear even more efficiently than before. The place was already polluted and strange.

But what brought people to select this part of the landscape for waste disposal in the first place?

Places that invite discards occur in various parts of the landscape. Marshes, deserts, woods, former mines, or quarries represent just the most obvious examples. These places are supposed to contain and control threats to more populated areas, primarily cities (Zapata Campos 2013: 50). In applied research, formal approaches have been developed to identify the places that would minimize the negative environmental, economic, social, and political consequences of waste disposal (Eiselt 2007). The formalization of the process is supposed to cover a range of variables and increase the overall credibility of the selection process. However, the selection of waste disposal sites can rarely be reduced to a mere application of the recommendations produced by research. There are more subtle factors that are difficult to account for in formal models. These factors include covert interests of stakeholders, informal negotiations, and, most importantly, classificatory logic tied to historical processes that pulled together certain entities to share a common destiny. It is exactly like Pavla Horáková's observation that I quoted above. In her novel about various modalities of strangeness, she contemplates the reasons responsible for the spatial clustering of pawn shops, sex shops, bars with slot machines, and tiny convenience stores, which all generate a sense of not fitting into normative expectations concerning neat urban space. They seem to be placed close to each other because of their shared position within an imaginary classification system.

Classification enables us to draw boundaries and put together events, things, places, and beings that are alike and differentiate them from those which are not. It means that the entities that are grouped together share similar attributes (see Cohen and Lefebvre 2005: 9). While cognitive scientists put most emphasis on experience, stimuli, and language to understand how humans categorize the world, this chapter is concerned with the spatiotemporal proximity of entities as evidence of a structure of ideas and practices. This proximity emerges via placement. It is guided by the "invisible forces of categories" (Bowker and Star 1999: 5) that can be clearly felt. One of Bowker and Star's contributions is that classifications do some work beyond the mere act of placing a thing into a box. Classifications literally change lives because they are spaces of significant political and ethical work. Certain things are supposed to be promoted, while others are silenced or avoided.

Mary Douglas (2002: 196) observed that avoidance is intimately tied to classification. Avoidance of the strange and polluting reduces the danger that the world will fall apart. As Max Liboiron and Josh Lepawsky emphasize, "any system must find ways to discard people, places, and things that threaten its order" (Liboiron and Lepawsky 2022: 61). Waste is part of this larger story. Amita Baviskar and Vinay Gidwani suggest that the negative charge of waste "... lies not only in its intrinsic physical or chemical properties, but is inseparable from where it is located in the spatial flows that form the social ecology of production and consumption" (2019). While the followers of Douglas—although not always in good agreement with Douglas's original intentions (see Liboiron 2022)—often emphasize that all that threatens is, in a certain time of its representational life, out of place, I am rather interested in where all that threatens should be placed to be separated from the rest. For example, mass waste in landfills is exactly in place because it is supposed to be there unless one prefers other means of waste disposal.

This approach means exploring where the already recognized category of impure, dangerous, or anomalous, which I prefer to label collectively as strange, finds its place. Although I agree with Douglas's core argument, I find her selection of the words dirt and purity unfortunate. Her argument about the danger that threatens order seems more general and goes beyond the connotations associated with dirt and cleaning. Hence, I prefer to use the word strangeness, which accounts for anything that is difficult to know, understand, classify, approach, or present and that may generate uneasy feelings. It seems closest to the Czech vernacular understanding of the word *podivnost*, which also covers a wide range of meanings comparable to strangeness. This is useful for thinking about entities and processes of diverse kinds that share the capacity to irritate or disturb.[1]

Placement and separation are critical steps in waste's disappearance. It can be seen as a way to limit the sphere of the possible. As Joshua Reno, building upon Bateson's and Deacon's ideas about entropy, says: "... mass waste disposal allows us to actively avoid alternative possibilities we would rather not experience or imagine ..." (2016a: 10–11). The same holds for other kinds of strangeness, such as military areas, minefields, industrial plants, animal farms, or virological labs that have the potential to increase the range of possibilities beyond the limits of expectation or even imagination. Spatial separation and avoidance of the entities with a capacity for indeterminacy (see Alexander and Sanchez 2018) is, therefore, part of the quest to live a meaningful life. Indeterminacy provides a potent vehicle for the elucidation of historical configurations emerging from human interactions with landscapes (Henig 2019). In this perspective, indeterminacy also includes experiential and affective registers that guide avoidance. In the Czech context, a recent volume edited by Radan Haluzík (2020) develops a closely related concept of *vague terrain* that captures uncanny urban places. The vagueness of these places stems from their interstitial nature and emptiness that, however, invites unexpected combinations of things, beings, or practices. Vagueness seems to capture part of the spectrum that the more general and versatile concept of indeterminacy refers to.

Avoidance encourages the disposal of the uncanny or strange through the management of presence and absence. As Kevin Hetherington says: "In engaging in such acts [i.e., disposal], members of a society can be seen to get rid of unwanted things and thereby to stabilize social categories of membership and belonging around their removal" (2004: 161). For him, disposal enables placement of absences, which are, however, never fully gotten rid of, and their indeterminacy lurks around. Reno (2016a: 161) extends the attention paid to absence to the concept of emptiness that may invite waste dumping. However, places that seem to be empty from a specific point of view might have a rich history that attracts various kinds of strangeness to join in the quest to become absent. Some places can be quite full, imbued with potentialities that serve as a magnet for making associations (Henig 2020). They follow the logic of a Czech proverb that "a crow sits next to another crow" (*vrána k vráně sedá*). In other words, a logic of placement, or displacement, of absences in Hetherington's sense is often contingent upon the presence of strangeness that attracts its various peers like a magnet. This seemingly paradoxical logic uncovers that presence and absence are intimately tied (see Sørensen 2010). To make something strange absent simultaneously means to make it present in a different regime of representation, place, or time. The creation and perpetuation of places that seem strange is often tied to an act of giving up. It is an attempt to concentrate the uncanny in *sacrifice zones* to sustain life elsewhere (Juskus 2023).

Understanding the placement of waste disposal sites receives a new spin when one pays attention to their history. Kathleen Millar (2012, 170) describes that the garbage dump near Rio de Janeiro was established in an area that was originally perceived as uncivilized, violent, and poor. Large-scale industrialization, destruction of the forests, and immigration from the poorer Brazilian states built upon this history and made the area absent from the social geography of the city (Millar 2018: 19). Reno's (2016a: 54, 141) research in Michigan showed the association between swamps, which were historically believed to cause diseases, the presence of poor and stigmatized people, and waste disposal. Famous Fresh Kills landfill, established in Staten Island's marshes, was preceded by a marine hospital and a waste reduction plant that set precedents for the landfill (Melosi 2016: 61). In Slovakia, the landfill near Košice was placed near stigmatized Roma communities (Filčák and Ficeri 2021), while the reversed logic of locating replacement housing for the Roma on the edge of a municipal landfill was applied in Turčany (Škobla and Filčák 2023). In Romanian Pata-Rât landfill, the emergence of an informal Romani settlement at its edge resulted from a combination of evictions from the city, racial prejudices, unemployment, and opportunities for cheap informal labor at the landfill (Dohotaru 2013: 194).

The link between waste and marginalized people is part of a larger process of racialization of labor in contemporary capitalism responding to environmental challenges (Dunajeva and Kostka 2021; Resnick 2021, 2025). Using examples of three Czech landfills, I examine why certain parts of the landscape perpetuate a sense of strangeness.

I pay special attention to history to understand how long-term practices, economic interests, specific events, material indeterminacies, multispecies encounters, and classificatory exercises took part in the transformation of these places into loci of strangeness. Using a metaphor of magnetism, I refocus our attention on the capacity of strange entities and activities to attract each other along a spatiotemporal continuum. As Gastón Gordillo (2014: 37) suggests, *magnetism* can be used as a metaphor for capturing a capacity of spatially situated material references to the past to attract new relations. By analyzing these relations, new histories, which contrast with the official accounts and history, may be unearthed. I will elaborate on the metaphor of magnetism in three senses. First, magnetism is not only about attracting but also repulsion. Magnetism keeps certain things away and encourages avoidance. Second, magnetism accounts for a certain regularity (attraction among strange entities in this case) but the pattern is not entirely predictable. In other words, there is space for attracting unexpected combinations of entities and activities that may erode an official version of history and create effects that are rather uncanny (cf. Stewart 2017). Third, magnetism itself is invisible, but its consequences can be sensed. It is reminiscent of other invisible agencies, such as witchcraft (Evans-Pritchard 1937), ghosts and spirits (Brož 2024; Tsing et al. 2017), and radiation (Hecht 2014; Jackson 2011), that do not have to be seen, but their effects can be serious. Using three case studies, I argue that strangeness sticks to places and tends to perpetuate itself through a series of magnetic relations over time. The logic stems from the human propensity to classify and dispose of entities whose open-endedness is dangerous and has to be placed away, as well as the continuities of historical processes. While disposal is often imagined as a management of absence, I point to a dialectical relationship to the opposite category of presence as a critical source of magnetism.

LET US STICK WITH THE POLLUTED PLACE

A man in his thirties in business attire is standing at a landfill. The wind is playing with his hair and tie the same way it lifts and releases pieces of paper and plastic that keep hopping across the landfill's surface behind the man. He looks into a camera, getting ready for his speech. He introduces himself as a technical director of the energy company called Warm, which runs the landfill. He points to garbage with disgust and says: "So, you have this in your backyard. Do you really want this?" Then he introduces a new solution that is supposed to be much better: ZEVO. The word is an acronym of *Zařízení pro Energetické Využití Odpadu* (waste to energy plant). He describes this cutting-edge technology with all its superlatives, including the production of heat and electricity for the nearby villages and the city. He does not forget to mention the word recycling and skillfully creates a vision of the apparently natural technological synergy between incineration and recycling. It is a no-brainer. Who might doubt the progress from the ugly and bad-smelling landfill to the slick, high-tech incinerator that can make garbage disappear entirely in a purifying fire?

I first saw this online video in 2013 when I searched for available information about Warm and its business. I knew that negotiations concerning the construction of the new incinerator right next to the Pureland landfill were in motion. The company was eager to proceed because the referendum in a nearby village of Lhota supported the construction. However, not everybody was happy. Environmental activists mobilized resistance. Two organizations, *Děti Země* (Children of the Earth) and *Sdružení pro udržitelný rozvoj* (Association for Sustainable Development), filed a lawsuit against a regional authority for bias in its decision regarding the zoning plan. The argument was based on a clash of interests. The regional authority issuing the zoning plan was, in fact, subordinate to the political representatives of the city, which owned Warm at that time and, therefore, were in the position of investors (Kaiserová and Mertl 2016: 23). After a series of law battles, Warm finally received all the permits, finished the construction, and put the incinerator into trial operation in August 2016.

The genealogy of the process deserves attention. At the turn of the millennium, the first discussions about the incinerator emerged. In 2002, Warm received seedcorn money of US$250,000 from the US embassy to cover a study evaluating the possibility of constructing an incinerator in the region (Kaiserová and Mertl 2016: 19). It was only one year later that Warm took over the already existing Pureland landfill, apparently with the idea that it could be a good place for constructing the future incinerator. From 2003 to 2017, the landfill served as a temporary side business during the transition to incineration. There are several indices that support this view. I recall a statement from a landfill manager from another landfill in the region who, to my surprise, made an intriguing comment that the leaders from the Pureland landfill were not *skládkaři* (landfill guys). He seemed to contrast them with his team that specialized in waste management and especially landfilling. Warm was very different, in his view. It was a company in the energy sector that reacted to an increasing pressure to reduce the use of fossil fuels and recognized waste as a resource (see Alexander 2016; Hultman et al. 2021).

The manager made a fitting note that "waste would become the new coal of the future." Therefore, it was not hard to understand why the heating company had become interested in waste management. Also, it was apparent during my fieldwork that the headquarters did not care much about the landfill itself as long as it did not cause trouble or generate fines for inappropriate conduct or environmental pollution. I do not recall meeting any of the top managers at the landfill or even hearing the workers commenting on their visits over the years. They often expressed a feeling of being forgotten by Warm and they would certainly make a lot of ironic comments on the managers who would come to see them and their landfill. In contrast, when the construction of the incinerator started in 2013, the managers and local politicians kept visiting the construction site to see the progress (Figure 5.1). The degree of interest in the new infrastructure was markedly different.

The Pureland landfill is located in the area of a former sand quarry, which was founded in 1954 to supply surrounding villages and the city of Pilsen with sand for construction purposes. During the 1980s, the quarry was slowly transformed into a place for dumping waste. In 1994, a new sanitary section of the landfill was opened to fulfill the requirements specified. The

Figure 5.1. Construction of the incinerator next to the Pureland landfill.

current owner of the land, the Voluntary Confederation of Municipalities, leased the landfill to the company Warm in 2004, which has run the landfill since then. After the official shutdown of the site for the disposal of municipal solid waste, the landfill has served to dispose of the bottom ash produced by the incinerator run by Warm. During my periodical visits to the closed landfill and the new incinerator between 2017 and 2022, Warm hesitated to close the landfill's tipping zone entirely. They were approaching the upper capacity limit for disposal, but it was clear that they wanted to prolong the landfill's afterlife as long as possible. One never knows what kind of polluting matter would need to disappear there in the future.

The history of the place indicates that multiple kinds of strangeness kept attracting each other over time. Once the quarry was abandoned and the subsequent expansion of wilderness turned it into an uncanny place covered with vegetation, people started using the place for disposal. During the period of state socialism, there were thousands of informal dumps all over the country. The former quarry was an apparent opportunity for dumping. When Warm took over the landfill, it was not because of the company's interest in waste disposal but because of the recognition that such a strange place can invite yet another polluting infrastructure. The place attracted the incinerator. Naturally, people complained about the landfill from the very beginning. The inhabitants kept mentioning especially an "incredible stench" and uncertainty about what was really disposed of there, given the stench. Some of them expressed their concerns with the "rusty fluid" (i.e., landfill leachate) that could be seen through the fences, and nobody knew whether it could leak and flow into the tiny village situated down the stream of a creek. Although there was resistance when the plans to build the incinerator were announced, it was not strong enough to stop the project. Many people were already accustomed to the presence of the landfill and the strangeness of that part of the landscape. The incinerator simply fitted into the already existing category of a place that can be sacrificed.

EXPLOSIONS AND DECAY IN THE WILD

When I conducted my first trip to the landfill called Vrch in autumn 2014, I was stunned. After I crawled through a hardly passable forest full of thorny bushes and dense spruce trees, the enormity of the hill emerging in front of me simply took my breath away. It was hard to believe that this one-kilometer long, half-a-kilometer wide and forty-meter high structure had not only been created by humans but also hidden so skillfully that most people have not been aware of its existence despite passing it on a busy main road located just 300 meters away. More interestingly, a pristine forest reminiscent of those from the Czech national parks made me wonder why the landfill was placed just here.

The history of the area where the Vrch landfill was established in 1996 can be traced to the eighteenth century. According to a local historian,[2] the Austro-Hungarian army established military crew in the nearby town of Rybíz in 1729. After a significant growth of the military presence in the town where the barracks for the dragoons and their horses was built, it was necessary to find a place for exercising. Therefore, 60 hectares of the land in the woods were assigned to the army to be used as a training ground. This military area is clearly visible on the map of the Second military mapping in the mid-nineteenth century. When Czechoslovakia was founded, the area continued to serve military purposes. After World War II, it was used as a firing range and impact area for air bombs and missiles. In the early 1990s, after the Velvet Revolution, the Czech army returned the land to the town of Rybíz, which was supposed to find a meaningful future for the disturbed landscape in the middle of the woods. There were two entrepreneurs who saw an opportunity to mediate the relationship between the representatives of the army and the Rybíz town council. There was another former military area in the region, which was polluted and more visible because of its close proximity to a nearby large city, and the army was supposed to clean it. As Pavel, a regional expert on environmental impact of infrastructural projects including landfills, says: "It was smart. They helped one group of soldiers to move their waste to another group of soldiers." The entrepreneurs found a solution in mediating the move of military waste to Vrch, where various remains of military presence were already present. They had to, however, turn the area of Vrch into a sanitary landfill to abide by a new waste act, which was passed in 1994.

The chronicle of Rybíz describes that in 1995 the town council approved an agreement about the future contract with a private company, ORI, concerning the construction of a landfill for municipal solid waste in the former military area of Vrch. The contract was signed despite a strong disagreement among the inhabitants of the town who expressed their concerns in a questionnaire survey, which seems to be a common pattern during the constructions of waste disposal sites (see Reno 2016a: 159). Multiple indicia suggest that there was pressure from the top to transform the problematic part of the landscape into a commercial landfill. The CEO of the waste company started publishing articles in the town newspaper in 1995, and "waste-promoting" texts kept appearing till 1998 when the process of the transformation of the former military area finished. The quality of these articles clearly surpasses others in the local newspaper. The author, an engineer by education, describes the evolution of waste management since prehistory and carefully crafts his argument for the necessity of building a new landfill near Rybíz. His playful linguistic style and creative use of examples and comparisons (e.g., "Middle Ages are ending right now!") make his texts both amusing and persuasive. He emphasizes the strict rules in waste management and modern technologies that prevent any harmful environmental impact. His articles were followed by a series of shorter texts by other authors who also promoted the idea of the new landfill. These were invitations to visit another landfill run by ORI to see its quality and safety, descriptions of the unacceptable nature of Rybíz's current non-sanitary landfill that should be closed as soon as possible, and an article of a pupil from the elementary school who pointed at the dangers of non-sanitary landfills. The pressing need for change is embodied in the calls, sometimes even dramatic, for collective responsibility to avoid informal waste disposal and, therefore, support the new landfill ("We are all sitting on the same branch, and some of us have saws in their hands.") All these messages promoting the change in the perception of the proposed landfill in Vrch were supported by special collection days when ORI waste lorries arrived at Rybíz and offered a collection of "the problematic kinds of waste" such as batteries, fluorescent lamps, oils, etc. to everybody for free.

ORI was slowly building its reputation and dulling the signs of aversion to the landfill. In June 1997, the new landfill at Vrch was officially opened to invite the first loads of its assumed capacity, four million cubic meters of solid waste. Interestingly, two months before the official opening of the landfill, a local newspaper published a brief article about Prime Minister Václav Klaus buying a summer villa in Rybíz from the mayor and accompanied the text with a photo of this top politician. This news turned life in Rybíz upside down. It drove an exceptional interest and excitement among the local inhabitants, who started bombarding the newspaper with their reactions. Once the landfill was opened, the inhabitants learned from the newspaper that the info about Václav Klaus was a joke. The mayor himself claimed that he wore a plastic mask of the prime minister and fabricated the story to amuse the people in the town. The timing of this event, however, raises suspicion. The debate completely diverted public attention from the official opening of the landfill to the topic that seemed much more exciting. Even today, questions about the creation of the Vrch landfill result in silence or lack of memory among the municipal officers. When ORI bought the land beneath the landfill in 1998, the old military era finished, and the landfill started building its reputation as a place with a voracious appetite for new waste and surprising tolerance for its origin or nature.[3]

Until May 2021, the remains of military activities had still been visible in the landscape before the construction works associated with the extension of the landfill site leveled them with bulldozers. Massive earthworks

Figure 5.2. Historical military earthwork at Vrch with the current landfill in the background. © Daniel Sosna.

with the top reinforced with rotten wooden planks and iron posts remain as silent witnesses of the times when Vrch experienced shooting and explosions (Figure 5.2). Surprisingly few people, including local woodmen, know anything about these earthworks hidden in a dense forest next to the landfill. It was a combination of patience and luck that allowed me to meet a 72-year-old man on a dusty forest road next to the landfill. This fairytale-like bandit wearing wellingtons and a funny felt hat, holding a stick with a sword-like handle, turned out to be a person who used to shepherd cows in the military area of Vrch during his childhood. He remembered not only the earthworks but also the changes during the last half a century:

> Not only the earthworks but there was also a bunker there. You cannot see it anymore though ... Soldiers used to train shooting from tanks' machine guns into the moving targets. They had the metal carts and tracks over there and let the carts go downhill while they were shooting. Also, they were training how to throw the grenades. But they were not here all the time. So, during the calm periods, we, the kids, could use the carts for fun. We were removing bullets from the unused cartridges and used slingshots [*šíbry*] to shoot on the targets. That was fun. ... But when the army left, I don't know twenty-five years ago or so, people stole everything ... all the tracks and carts, even metal stairs and chimney from the building. One guy still has a shelter for the tanks in his backyard. Remaining military rubbish [*bordel*] and various debris that nobody wanted were simply left here ... During early 1990s somebody even tried to grow cabbage here for one or two seasons. But at the end, they made this landfill here ... Oh man, don't forget, the iron posts on the earthworks are waiting there for you![14]

The old-timer suggests that strangeness, while often having negative connotations, has another side as well. It is like an unexpected case of insects and crabs floating on the plastic debris in the Pacific Ocean, taking advantage of the strange accumulations of plastic waste to thrive (Strain 2012). The former military area was not just a dangerous and polluted place, but it generated positive engagement and affect. The people living near the military area were not only disturbed by noise and explosions, but they also used the place for grazing animals, searching for mushrooms, children's games, and were even able to collect and sell or reuse various kinds of unwanted military stuff once the army left. Military waste offers a potential for value creation and transformation along various registers (Henig 2012, 2019; Kim 2016; Reno 2020; Zani 2019). It can be harvested and sold for its exchange value (Kim 2016: 178), enable the generation of surplus value through rentiership, or can be displayed in museums to show its value as a material carrier of identity and authenticity (Reno 2020: 55, 99). Moreover, it can be just the presence of military waste that enables new engagements in its surroundings via the activities that generate value (Henig 2019: 106; Reno 2020: 102) or imbue places in the landscape with ecological value (Kim 2016: 167). In Vrch, value was emerging through the activities of the domesticated animals fed on the meadows and people collecting mushrooms and berries in the area, which was not officially accessible. Various kinds of army equipment and weapons were revalued in children's games as a source of adventure and fun. The collection and subsequent sale or reuse of material objects left in the military area after its abandonment was well comparable to the activities of bootleg miners vividly described by Rakowski (2016: 125–26). The miners were extracting various materials from the former mines within the grey zone of confusion about the nature of property rights. The same was true when the army retreated from Vrch. The discordance between the formal property rights and the obvious lack of care and start of ruination at Vrch led local inhabitants to collect and transport away whatever seemed valuable.

I would, however, extend Henig's (2019) argument about the capacity of military presence to generate unexpected engagements and meaning further. Here, it is a wider range of abandoned material remnants of military presence, such as rods, rails, bunkers, concrete blocks, and earthworks, that mobilize relations and action, including those that grow from the indeterminacy of the place. Moreover, it is not only the materiality of military waste per se but also the various consequences of military action that contributed to the indeterminacy of Vrch. Indeed, regular explosions producing depressions on the ground, forest clearing that changed conditions of light or humidity, and seemingly destructive movement of the military equipment created environmental conditions suitable for rare forms of life such as endemic plants or insects. As Tsing (2015: 160) argues, building upon the work of ecologists, disturbances are not always bad. Their nature is contested. "Disturbance is never a matter of 'yes' or 'no'; disturbance refers to an open-ended range of unsettling phenomena" (Tsing 2015: 161). While disturbances could pose a problem for the foresters, the municipal officers, or shy mammals, it certainly was not a problem for rare orchids and tadpole shrimps thriving because of the periodic explosions creating temporary water pools and marshes. Therefore, the strangeness of Vrch entangled with its indeterminate nature was contingent upon a wider range of factors than the materiality of military waste only.

Despite the apparently rich life of the former military area, its official history is largely denied. There are no information panels or trails that would tell the visitor about the history of the place, although similar projects have been mushrooming with the support from the state and EU during the last decade (see Woitsch and Pauknerová 2019). There are at least three such trails in the landscape cared for by Rybíz. Only the earthworks and a few concrete blocks in the deep woods suggest that there was something other than the landfill before. The strangeness of the place is evidenced in a tendency to make it absent and forget its history. Although there are some people who still remember the pre-landfill times, notes about Vrch in the chronicle and two official historical books about the town of Rybíz and its surrounding are surprisingly scarce. My questions relating to the history of the place were responded to either with silence or shrugging shoulders among the municipal officers. The historian who wrote an entire book about the town and worked for the municipality for thirty years told me that he knew so little about the Vrch area that it was not worth a meeting unless I would change the topic of my interest.[5] The place has become absent not only from a social geography in Millar's (2018: 19) sense but also from official history and collective memory. Forgetting is an inevitable part of relations to the landscape (Pauknerová 2019).

All the available information suggests that Vrch is a kind of place that should be avoided. Douglas argues that avoidance is a form of public recognition of the boundaries that delimit a structure of ideas (2002[1966]: 196). The strangeness of Vrch reminds everybody that some things and processes should be kept aside to keep the world meaningful. This categorical avoidance, nonetheless, creates interesting economic opportunities for informality at different hierarchical levels ranging from top managers to "hyenas" at the bottom (Mašek 2023). As Ledeneva puts it: "They are [i.e., informal practices] often invisible, resist articulation and measurement, and hide behind paradoxes, unwritten rules and open secrets" (2018: 7). An inability to see the processes at the landfill gives the managers and workers space for dumping inappropriate materials, covering leaks of landfill leachate and its inappropriate treatment, altering statistics, and informal collection and transport of materials out of the landfill. Normal people do not visit the area much, which makes the surroundings of the current landfill a wonderful place for "nature" to thrive. Wildlife flourishes in the woods because its most lethal enemy, humans, do not cast their step there much. It has been only the specialists such as soldiers, landfill workers, lorry drivers, and foresters who were officially able to cross the border and engage with the strangeness dwelling inside. Children, mushroom pickers, or scrap collectors were entering only when they transgressed the rules.

Birds play a special role in the multispecies thriving in the Vrch area. Ravens, in particular, have the capacity to retrieve food scraps from the surface of the landfill and transport them to the surrounding woods. Emerging hot spots filled with hundreds of diverse plastic containers for food or drink, which lie scattered across the forest floor, embody evidence of not only ravens' feasting but also their capacity to create new ecologies enabled by waste. Remainders of food attract wild boar, foxes, feral cats, deer, mouflons, hawks, and kites to finish what the ravens did not. Ravens orchestrate these activities and relations of *marginal gain-making* (Hoag, Bertoni, and Bubandt 2018) that demonstrate the potential of multispecies encounters to resist the disappearance of garbage within the flows of the official waste stream. Moreover, the eerie aesthetics of these hot spots in the woods magnify the sense of strangeness of the entire area.

Military activities that transformed Vrch into a strange place, nonetheless, were not the only source of magnetism for the construction of the landfill. It was another usual suspect of pollution and avoidance known from anthropological theorizing: the pig (Douglas 2002[1966]; Harris 2001[1984]). About half a kilometer from the future landfill there was already a large-scale pig farm whose history went back to socialist times. In 1992, a new company was founded, which created a massive farm for 29,000 pigs. Such a farm does smell, despite the existence of a sophisticated digester that turns animal feces into electricity. As a local librarian from Rybíz told me: "You don't wanna be around when they release gas from the digester. Who knows why they do it every now and then, but it is really terrible. We must close the windows to escape it."

Regardless of these periodic releases from the digestor, odor tends to surround Vrch. It is not only when one walks around. The odor is present every time one drives on a busy road that goes between the pig farm and the landfill. For local inhabitants, olfactory experience is part of space ordering when a smellscape consists of specific odors related indexically to specific places (Jackson 2011). Interestingly, the quality of odor being dynamic and "neither a thing of earth nor of sky" (Keil 2021: 9) may create ambiguity. In the Rybíz region, one never knows whether the odor comes from the farm or the landfill. In fact, this olfactory ambiguity serves as a tool to impede a potential critique from the people living not far away or just passing by. Anyway, the pig farm served as another magnetic attractor for the landfill to join the mutual quest of keeping the continuity of the strange place.

In 2019, my suspicion about the magnetic properties of strangeness received unexpected proof. Pavel, the regional expert on construction projects, mentioned that there is a plan to build a new shooting range for the police, army, and athletes right next to the landfill. This time, however, there was a problem with the endangered species of plants such as Siberian iris (*Iris sibirica*) and lesser-butterfly orchid (*Platanthera bifolia*) that grew in the clearings close to the landfill. Pavel took part in the preparation of the report concerning the environmental impact and was uneasy about it. He wanted to protect the plants but was well aware of facing a powerful group of people from the county council, Czech Olympic Council, the mayor, representatives of the regional police and army, as well as heavy hitters from the construction and waste business. In spring 2020, when I saw the bulldozers clearing the forest next to the landfill, I knew that the permission had been granted. The cunning entrepreneurs killed three birds with one stone. They had to dispose of massive amounts of soil and debris from the construction of a new expressway that would bisect the woods between the landfill and the pig farm and, at the same time, use the unwanted soil and debris for the construction of earthworks that would isolate the firing range from its surrounding. The ORI company that runs the Vrch landfill would be allowed to contribute to the construction of the earthworks with their "soil" from the landfill to empty it and increase its capacity for deposition. It was extremely difficult to stop such a brilliant project where the multiple companies that build the firing range, expressway, and run the landfill, would benefit from it. At the same time, Rybíz would receive payments for the deposition of "construction waste" on its land, and the athletes, police, and army officers would have a place for training. Indeed, the facility is supposed to host prestigious events such as international shooting championships.[6] It does not get any better for the actors promoting the project, and loss of a few "flowers" or noise from shooting cannot change it. Moreover, based on my long-term experience with waste management in other landfills, there is great potential for both formal and informal altering of the size of the earthworks and the material that can be deposited there. The idea of a new shooting range is so great for so many powerful actors that the unwillingness of the official representatives to talk about the history of Vrch makes perfect sense.

I argue, however, that none of the projects that have been transforming the landscape at Vrch since the 1990s would have happened if Vrch had not been already perceived as a strange place. Training and shooting by Austrian-Hungarian dragoons, Czechoslovak tank crews, and pilots dropping the bombs on targets, abandoned ruins after the withdrawal of the army, pigs, including their smell in the nearby farm, all created an uncanny aura that served as a magnetic attractor for the more recent use of the place for landfilling and shooting. Reno (2016a: 161) argues that emptiness can function as a prerequisite for waste disposal and the creation of escape. While there was certainly a sense of emptiness, imagined as an opportunity for value creation through training of people and generation of profit or prestige, Vrch was never really empty. It was the presence of strangeness deeply ingrained in the images of the place that enabled its various kinds to stick together and create a continuity of the weird place that has been fading from the official registers of representation despite its occasional appearance in times of rapid change.

Hirsch (1995) suggests that landscape is best understood as an interplay between the "foreground" and "background" of social life. For him, the landscape offers a view of "... an ordinary, workaday life and an ideal imagined existence vaguely connected to, but still separate from, that of the everyday" (Hirsch 1995: 3). Viewing Vrch through this perspective reveals forgetting and a lack of reflection on one hand and indeterminacy of imagination on the other. The everyday dimension was hidden from sight for a long time when the military area and later the landfill were separated from its surroundings by a dense forest, allowing it to perform activities and manipulate materials that might make external actors feel uneasy. The foreground remained largely unreflected and silenced despite being quite obvious in terms of the everyday rush, noise, and smell. At the same time, the background dimension of Vrch lent it an aura of long-term strangeness. It could be a place of unfortunate injuries from combat, environmental disturbance and pollution, strange smells produced by unknown substances, disappearance of anything imaginable, treasures brought to the landfill by garbage lorries, exceptional mushrooms for their pickers, animal trophies for the hunters waiting in ambush for the animals breaking the fences to feed themselves on trashed food, or a place where one can generate unexpected profit. The background was more critical for the long-term continuity of the place because it always conveyed the feel of strangeness that lured its followers to "inhabit" the place. An explicit link to death or dying, however, has never been articulated. The following case from Mýtina will show that waste and death can go hand in hand.

BURIALS WITHOUT GRAVESTONES

We are sitting in a garden, and Mr. Smrk describes a story from the end of World War II in Bohemia, which indicates again that different kinds of strangeness attract each other. The story is so intriguing that I present its excerpt here:

> It was May 1945 and Germans tried to escape the punishment. There was a demarcation line here and our town was divided into two halves. You know, the Russians were over there where the current town hall is and the Americans were on the other side, if you know where the pub is these days. And there was a land of nobody in the middle, which dissected the town ... [I omit the part describing Russian and American soldiers] ... I do not know how everything started but my uncle Pepa told me that they caught them [i.e., the Germans] in Nová Paseka in the woods. There were five of them and when they brought them on a cart to this little square in front of the church [showing the location of the square with his hand], they were heavily beaten ... actually mostly unconscious. On the way, people were beating them again and again ... even women were using hoes to hurt the poor guys. They were still alive but really trashed. People put them on the ground and the local broadcast was inviting everybody in town to come and see the humiliated SS men [*esesáky*] and have an opportunity to kick them [*kopnout si do nich*]. One of the Germans looked like a sixteen-year old soldier, but who knows who these people really were. The farmer Vlasák led the group and the local policeman Hrom participated as well. When the public beating by local townspeople finished, the group loaded the cart with the Germans and took them to the SS woods [*esesácký lesík*], which was situated exactly on the demarcation line. I do not know how they finished them, but I heard that Hrom, because he had a rifle, shot all of them and people, who were there, buried them in the SS woods, which is right next to the current landfill as I showed you. Interestingly, candles kept appearing there since that time.

I first heard a small fragment of the story about "the Germans buried beneath the landfill" in 2016 from a historian who had heard some rumors and encouraged me to explore the topic. It took a long time to find people who would know anything about the event and, most importantly, were willing to talk about it. Although the event had to be experienced by many inhabitants of the town, it happened more than 70 years ago. Most witnesses had already died, and those who witnessed or heard about it from their older relatives were not particularly proud of this part of their history. There is no entry about it in the town's chronicle. Indeed, most people that I met and talked to in the town had no idea what *esesácký lesík* or *u esmanů* (another version of the place name) meant. A local enthusiast and doctor of natural sciences helped me initially because he had created an educational trail in the woods and examined historical sources. Although he himself was really surprised to hear the story,[7] he directed me to a local registrar who might know something. The registrar had heard a short story from one of her older relatives who was dead, but directed me to Mr. Smrk. The registrar presented the story as a narration that had to be understood with caution but she confirmed that she saw the candles in *esesácký lesík* when she was young. Interestingly, she remembered a detail from the narration of her relative that caught my attention. She heard that, during the lynching, somebody cut a hair braid from the head of a German girl. This little detail suggests that the labeling of the place of burial as related to the SS members and the entire framing of the story as revenge for the evilest members of the Nazi regime is a reflection of the need to legitimize the murder rather than the reflection of victims' identity.[8] Since place names can be used to convey moral messages (Basso 1996: 24), the reference to SS members can be understood as a way to emphasize the moral justification of revenge. As Mr. Smrk's wife suggested: "They were not SS members but regular soldiers." The notes about a sixteen-year-old soldier and a young girl fitted the picture of omnipresent uncertainty, anger, and eagerness for revenge at the end of the war, regardless of the guilt. Recent popular accounts such as Josef Urban's novel *Habermann's Mill*, Kateřina Tučková's novel *The Expulsion of Gerta Schnirch*, or a movie by Bohdan Sláma, *Shadow Country*, show the complexities of the expulsion of Germans at the end of the war and the wave of hate that hit various people regardless of their real collaboration with the Nazi regime.

Despite the lack of detailed information about the victims, a few more people from the town confirmed that the story had a real basis and candles were repeatedly appearing over the decades in the place of the assumed burial of the "Germans." One may speculate on who was bringing the candles, whether the girl was a Czech lover of a German soldier, or whether these "Germans" were even soldiers, or mere civilians who just wanted to escape from the Russian sphere of influence and cross the demarcation line to become prisoners of war on the American side. The more important point for my argument, however, is the spatial dimension of the story. The victims were brought from a tiny village in the middle of the woods where killing and burying would be much easier. Instead of that, they were exposed to public lynching and later killed and buried in the demarcation line. In fact, the demarcation line was not a line but a liminal zone between the two military powers. Such zones can be formally treated to abide by the rule of "no touch" but, at the same time, could be places of various activities, as Kim (2016) shows at the Korean Demilitarized Zone. In Mýtina, local inhabitants apparently could enter the zone and took advantage of its special status. Murdering people in that zone meant that neither Russian nor American laws would operate there, or at least they could be contested.

The demarcation zone, nonetheless, was not the only source of strangeness in a categorical sense. The region became the center of coal mining in the mid-nineteenth century, and some surface mines were already abandoned or close to shutdown. The mine in the vicinity of *esesácký lesík* was closed in 1935. It means that this recently abandoned industrial landscape consisted of various slag heaps, depressions, and ruins. It was in the process of transformation from an industrial place into an abandoned place, inviting various forms of life, among which my interlocutors emphasized bushes and coniferous trees, to colonize it (cf. Tsing 2015). Rubble and remnants of mining did not turn the place into a worthless part of the landscape. In contrast, they served as attractors of future possibilities. As Gordillo argues: "Rubble exerts positive pressure on human practice and is constitutive of the spatiality of living places" (2014: 11). I understand his call to turn our attention from a fetishization of ruins to the multilayered potentiality of rubble as a way to recognize and acknowledge that rubble invites new relations and engagements, which are not reflected in the official accounts. Abandonment of the mines in Mýtina resulted in such new engagements, including the killing and burying of the "Germans" there. The strangeness of the place served as a magnet to attract the vendetta against the defeated enemies. It might be useful to note a striking similarity in the treatment of the enemies by the Nazis. During the war, the Gestapo set up a regional execution ground in a former quarry in one of the villages in the region.

The history received a new spin when the abandoned mine with the burial of murdered "Germans" became the dump site for the disposal of old industrial cables from the nearby engineering works. During state socialism (1948–1989), the place served for large-scale dumping of industrial waste, which was dangerous because of frequent spontaneous fires. Indeed, the place claimed another victim when a local fireman died after being overcome by toxic fumes during one of these fires on the dump. The place was also used for the informal disposal of household waste, tree stumps, and debris associated with the construction of a new highway. In 1993, it started functioning as an official sanitary landfill for municipal solid waste disposal. While there are several former mines in the area, it was the one in *esesácký lesík* that was selected for waste disposal. The landfill operated till 2019, and it is currently being recultivated. During my regular visits since 2016, I have noticed what I call *associative waste* in the vicinity of the landfill. This kind of waste sticks to places because of its association with official disposal. People unwilling to pay for their garbage disposal simply dispose of it near the fences that delimit the landfills.[9] While it is common in many landfills, the case of Mýtina is special. It was not just the landfill itself that attracted the associative waste but also the irregular terrain full of depressions and heaps of former industrial rubbish covered with trees and bushes that created opportunities for waste dumping. A degree of informal dumping has been exceptionally high, which was not only my impression based on movement in the landscape but also appeared in the discussions with my interlocutors, who have been aware of and often terrified by the amount of junk in the woods. Again, a mutual magnetism between different kinds of strangeness has produced a place that is difficult to classify and that continues to be strange despite various changes during the last century.

The history of Mýtina supports the view that humans and things are part of a general project of disposal despite their differences (Hetherington 2004).[10]

While acknowledging cultural diversity, Hetherington finds similarities in a two-stage process of getting rid of both humans and things, including the spatial management of their absence (2004: 168–70). Although his use of Hertz's (1960[1907]) theory of funeral rites downplays the problematic assumption about secondary mortuary practices taken as a general model of funerary rites, it demonstrates that there are striking similarities in the disposal of things and humans. This finding resonates with the studies pointing at a shared capacity of things and humans to be disposed (Bauman 2004) or similarities between recycling and ideas about death and regeneration of life (Graeber 2012). The most important point, nonetheless, is Hetherington's view of disposal as a mediation between the presence and absence of dead bodies and waste, which has social consequences. The history of Mýtina might be seen as a reversed version of Reno's (2016a: 217–19) call for resocialization and repersonalization of landfills that might be reached via burying the dead there. Reno searches for the "... ways to get closer to our sites of final disposal, to resist and reimagine their separation from our lives" (2016a: 223), regardless of how absurd this suggestion might sound to anybody outside the sphere of high-brow academic theorizing. As Reno says, landfills were originally designed to foreclose social relations between the waste producers and waste sites. The case of Mýtina demonstrates that the effort of foreclosing social relations via special separation involves a wider range of phenomena than just mass waste. It was the strangeness that co-produced the place where mining disrupted the landscape, the victims of war were killed and buried, and various kinds of rubble and garbage were deposited over decades. The burial prior to the emergence of the landfill served as a vehicle of foreclosure and separation rather than resocialization in Reno's sense. It is not only because of the reversed order when death and burial preceded the landfill but, more importantly, because of the nature of the death. From ethnography, we know that strange deaths such as drowning, murder, suicide, or being struck by lightning often require different treatment at burial (Ucko 1969). Such an abnormal death contributes to the strangeness of places that tend to be kept aside. Regardless of whether we ever reach the model of burials in landfills or not, there might still be burials that will require different treatment and placement to reflect the human capacity for differentiation and avoidance of entities that simply do not fit.

The spatial separation of the strange is not just an act of classification, but it has an economic dimension. Keeping strangeness out of sight enables the relevant actors to generate value. Waste management is a silent engine of national economies. Even in small countries such as Czechia, it is a flourishing industry that generates billions of euros every year and attracts international investors. Being powerful and out of sight also enables various kinds of informality such as manipulation of statistics, control sampling, capacity of deposition, shifts in classification, acceptance of problematic waste, and informal scavenging of the waste workers (Sosna et al. 2024). It is of no interest to anybody in the business to resocialize and engage with the public. Indeed, when the new waste laws were negotiated in parliament, they initially started in the secret regime before public pressure forced the politicians to change it. In addition, it is not only waste management that prefers to be hidden. The other kinds of strangeness described in the stories in this chapter, such as military areas, pig farms, mines, and quarries, tend to be kept aside and intimately tied to economic interests. A controversial case of Lety u Písku, where a concentration camp from World War II was turned into a pig farm, can serve as an instructive example (Polansky 2014). During the war, the camp served for the separation of Roma and Sinti, and exploitation of their labor before their extermination in Auschwitz II-Birkenau. In the early 1970s, a large pig farm was built there to continue the separation of impure beings, despite the attempts of the Romani association to create a memorial there. Since the late 1990s, multiple actors have tried to stop the piggery and establish a memorial, but it was not until 2018 that the government finally bought the farm from the prosperous commercial company to enable the creation of the memorial. It was an incredible period of two decades when the economic interests kept defending the strangeness of the place and drawing upon the mutual magnetism of impure people and animals despite intensive discussions and publicity. Strangeness simply sticks to places and has the power to perpetuate its existence in various forms. It embraces diverse kinds of strangeness that may include material objects (Gordillo 2014; Henig 2019; Pétursdóttir and Olsen 2014), human beings (Baviskar and Gidwani 2019; Dohotaru 2013), or strange animals attracted to disposal sites (Reno 2016a: 164). Their mutual entanglements perpetuate long-term categorical distinctions of certain places as being accumulations of whatever communities need to make absent.

When I carefully suggested the idea about the potential association between different kinds of strangeness to Pavel, my interlocutor, and expert on the environmental impact of large-scale infrastructural projects, he found it amusing. He was too friendly and kind to burst into an overt laugh, but it was clear that his training in natural sciences and long-term experience in applied research made him suspicious of humanistic interrogations of a self-evident sphere of practical reason. The magnetism I talked about was as invisible and absurd as witchcraft. For him, everything was obvious: waste disposal was determined by economic opportunities and the interests of the actors who recognized these opportunities. In this logic, any place can be turned into a landfill, firing range, or pig farm. Although I respect the power of calculative reason (see Gudeman 2016) and tried to acknowledge the economic dimension in the management of strangeness in this chapter, I insist that Pavel's straightforward explanation is insufficient to either explain why certain parts of the landscape became what they are or to understand what it means that they become what they are and what futures are waiting out there. Landfills, firing ranges, or pig farms do not tend to grow next to shopping malls, vineyards, or memorial sites. I will turn to classic wisdom of Sahlins, who insists that man (sic) "... must live in a material world, circumstances he shares with all organisms, but that he does so according to a meaningful scheme of his own devising ..." (1976a: viii) Although his critique of practical reason became rusty and one might criticize him for being andro- and anthropocentric, totalizing the concept of culture, or not leaving space for agency, the core of his argument that practical reason is simply not sufficient still holds.

This chapter builds upon Douglas's (2002) idea that humans tend to avoid the strange and this avoidance relates to the ideas about the nature of the world and its perpetuation. Although one might criticize Douglas for being too rigid, anthropocentric, or misunderstanding the nature of contemporary mass waste that does not lose its capacity to bite back, the gist of her argument provides a fruitful background for examining the relations between different kinds of strangeness because their effects can be sensed in the landscape. When Hetherington (2004) and Reno (2016a) engage with the concept of absence and develop a discussion about its spatial management, it creates space for further thinking. I demonstrated that avoidance of the strange—regardless of whether it is former mines with rubble, forests disturbed by military activities, dead bodies of war victims, pig farms with their specific aroma, dumps of industrial waste, or sanitary landfills—unfolded along the spatiotemporal continuum where both presence and absence were mobilized. While the strange is supposed to become absent, it never really disappears. Its indeterminacy may lead to its presence in a different place, time, or regime of representation. I argued that strangeness stuck to places and tended to perpetuate itself through a series of magnetic relations between the entities and activities, which are, in a certain sense, alike over time. Strangeness sticks to places. The magnetism between different kinds of strangeness, which operates in space and creates temporal continuity, has more strength than the people who engage with the strange are usually aware of or willing to acknowledge. The *invisibility* of this magnetism is reminiscent of magnetism in devices with electric engines. The effects of the magnetic field are evident when the engines convert electrical energy into mechanical energy and perform the work for which they were designed. The magnetic agency, however, is generally unrecognized or even unknown by the users of these devices. The same is true for the people who live within the magnetic fields of strangeness without noticing it.

This chapter is not intended to accept structuralist thought uncritically but to acknowledge that one of the ways humans apprehend the world is to classify it and organize action accordingly. This includes the acts of placement of entities and activities that make people feel uneasy because of their indeterminate nature. This unsettling openness may cause trouble because there are always possibilities that humans would rather not experience or imagine. The case of Lety u Písku, however, suggests that even the sticky strangeness can be eroded when extreme energy and patience are devoted to a transformation of the place from being a space of multispecies exploitation and forgetting into a space of reflection and remembering. It shows that the kind of magnetism described in this chapter always leaves space for the unexpected that may subvert the dominant magnetic attractions and change historical narratives and the place itself.

NOTES

1. My use of strangeness is only weakly linked to Simmel's concept of the stranger (*der Fremde*), which is much narrower and related to a person who has a specific position within society. Still, Simmel works with space. He points at an intriguing paradox of being simultaneously near and distant (Simmel and Wolff 1950). This is familiar for waste, which is essential for life but needs to be kept away. In a sense, the strangeness of waste is a reversed version of the same paradox. It is distant after placement while remaining intimately close (it carries signs of life).
2. I cannot cite the book of this historian because it would directly identify the town and landfill.
3. Several people, including lorry drivers and one of my long-term interlocutors, a manager of another landfill, told me that they were shocked at the amount of construction debris and soil deposited in the Vrch landfill as a "technical material" for the construction of landfill's structure. There is a strict limit for the proportion of this material at landfills, which was clearly ignored at Vrch.
4. The old-timer suggested that I could take advantage of the fact that I knew about the forgotten earthworks. He encouraged me to collect the heavy iron posts and earn money for taking them to a scrapyard.
5. This retired historian was in his eighties and was not willing to talk about Vrch. It was clear that he knew something but was afraid to share it with me. His entries about the life in the Rybíz region in the chronicle were incredibly meticulous and contrasted with a complete disregard with Vrch. I can only speculate about the reasons for this neglect, but he has been clearly avoiding the theme, both in his writing and personal interaction.
6. An avalanche of superlatives include the support from a Czech Olympic gold medalist in shooting, interest in training of "foreign intelligence agencies," and great logistical position near the highway.
7. The doctor moved to the town a decade ago and this story was hidden from him despite his projects to explore local history and raise awareness of the past among

the local people. Interestingly, the local chronicle did not know about the story either. Their surprise indicates how sensitive this issue was in the town.

8. The anger is not difficult to understand given the fact that just in the small region including Mýtina and three other small towns, 110 people were executed by Gestapo for insurgency during the war. The information comes from the official book published by a regional historian.
9. I must emphasize that this kind of waste is distinct from the plastic and paper bags that fly out of the landfills, or various food containers that animals transport to landfills' surroundings.
10. Humans and things might even become a single social being facing its death as a whole (Mašek 2023).

CONCLUSIONS

As I have demonstrated in this book, Czech society expresses a pressing need to make waste disappear. This need is not unique but resonates widely across space and time. Unlike other places, Czech garbage is carefully contained in bins and containers or hidden in underground chambers. Garbage must be placed in containers, lids must be closed, and subsequent transport to *nowheres* creates a sense that garbage ceased to exist via an act of disposal. The supposedly more efficient approaches to waste treatment rely on the circularity of economic models or the power of fire to dispose of waste. Technological promises and smart designs are supposed to bring us closer to the point of absolute disappearance, which is the ultimate goal. Imperfections, residues, and various underflows of waste streams are suppressed. This book's argument has been that the struggle for disappearance is neither realistic nor always well-targeted. It suppresses rich forms of living that unfold in intimate relationships with waste and deflects attention from the more serious challenges of reducing the destructive forms of life on the planet.

Waste finds a plethora of ways to reappear after it becomes part of underflows, whose presence can be discerned only when they approach the imaginary surface. The material agency of disposed garbage certainly has its limits and unfolds primarily in places that do not appear on maps. This does not mean, however, that garbage cannot participate in new relations. These places create affordances for diverse forms of life to flourish. I demonstrated that landfill workers, waste pickers, and various nonhuman actors creatively engage with garbage at landfills to recover value. Their salvage activities encompass a diverse range of items and materials, extending far beyond landfills, as their treasures are passed through the networks of relatives, friends, and business partners.

Moreover, the case of scrap metal indicates that these activities are far from being marginal. In times of problematic geopolitical access to and mining of primary resources, these activities prevent significant quantities of metals from being lost in massive accumulations of matter at landfills. Recovery of non-magnetic metals through sorting in technologically sophisticated infrastructures, such as incinerators, remains a challenge. The moral labor of the scavengers remains unrecognized or silenced, even though their activities fit perfectly with the political plans to achieve sustainability via saving materials and the prevention of waste. These people practice exactly what Joe Smith and Petr Jehlička (2013) call quiet sustainability. In addition to not being cool enough to become heard, salvaging activities at landfills are more than the self-provisioning of food that Smith and Jehlička originally had in mind. Scavenging and tinkering happen in a grey zone where normativity and legality become blurred. The fact that each waste picker and landfill worker I worked with recovered tons of scrap metal per year is not announced aloud. It occurs in silence, outside the public eye, thanks to the tolerance of some managers and police who share a moral conviction that salvaging discarded items is good. Salvaging things and materials from landfills resonates with the moral perception of gleaning as a traditional practice that helps to mitigate social inequality.

Giving new life to garbage is not only about recovering value tied to things and materials. The activities themselves have value. They represent highly creative engagements with matter that build upon previous generations of relatives developing knowledge and skills of finding, processing, transporting, and exchanging things that initially seemed valueless. The skills of disassembling devices without special tools, omnipresent improvisation, and patience were breathtaking, given their overall success rate. Their potential disappearance after landfilling ends represents a kind of loss that is not featured in the strategic documents outlining the future of waste management in Europe. It is an unrecognized collateral damage to the economy, social system, and cultural heritage.

Garbage at landfills has a strong semiotic presence despite all the attempts to make it disappear. As I argued, it is because landfills can represent a powerful synecdoche of all wastes. Due to this special potential, landfills can serve purposes beyond mere waste disposal. Attention, which has been directed toward landfills and their elimination, can lead to certain actions. The EU's politics of resource use have encouraged a shift away from wasteful lifestyles via an emphasis on waste prevention. A challenge arises, however, in that frugal life and waste minimization are not entirely compatible with consumer expectations and basic market assumptions. To resolve the trouble, one may deflect attention from the main causes of a potential crisis to the unpopular disposal at landfills that can stand for all our sins.

Turning disposal at landfills into a scapegoat is supposed to make the world meaningful again and secure continuity. The ongoing transformation of waste management in Czechia is more than just a shift from linear to circular imaginaries, technological advancements, and an attempt to reduce wastefulness and the creeping bodily and environmental harm embedded in slow violence. It seems to be a social mechanism that attempts to restore balance and sustain life via an act of giving up.

WHAT CAN BE DONE?

Thinking about garbage and judging its management's success in terms of its efficacy in disappearing would be a fallacy. As Joshua Reno (2016a) argued, the solution is not dreaming about the world without waste but cultivating our abilities to live well with it. Large accumulations of garbage have a strong reflexive potential that can remind us of the lives we live and the consequences that such ways of life bring. Therefore, letting garbage disappear without a trace would prevent us from recognizing the real trouble of our time: consumerism that has gotten out of control. Stories of discarded loads of intact gummy bears, unused sandpaper, and the amounts of iron and copper disposed of at landfills, which were presented in this book, capture the wasteful nature of contemporary life and the careless approach to resources. These stories are painful reminders that may prompt us to reconsider the false conviction that waste can disappear easily, and that one can forget about it once the garbage bin lid is closed.

Patrick O'Hare (2022a) argued that attempts to make waste disappear, either as a category through acts of reclassification or as matter that vanishes within enclosed infrastructures, harm vulnerable people whose lives depend on access to abandoned and discarded things. The increase in surveillance technologies and reinforcement of fences at landfills has lowered the chances of scavengers recovering things from garbage. Technological advancements in means of waste management tend to exacerbate this negative trend. Czech incinerators have magnetic devices to separate iron from the rest of the bottom ashes coming from the furnace. The automatic separation lines make it almost impossible for the informal activities of the scavengers to be performed. The companies control tons of collected scrap iron, and the money ends up in the pockets of the owners of these expensive infrastructures. The traditional laborers at landfills, who either lived off the collection of scrap metal or used it as a substantial extra income to supplement their modest earnings, remain empty-handed. The same holds for the circular economy plans that suffer under the dictate of the market. Opportunities for investors, startups, and other discard-related companies result in what O'Hare (2022a) calls another round of primitive accumulation of capital. New kinds of enclosures prevent access to waste as a resource and deteriorate the lives of people whose families have been dependent on access to waste commons.

When developing waste policies, broadening expert perspectives beyond usual technical or economic ones would be useful. If we acknowledge that the generation and management of waste are inherently social, it is crucial to understand the relationships to waste in which humans, as well as other beings, participate. The worst solution would be to view waste as an issue that can be solved by another innovative technology that would render waste disappear with greater efficacy. This book describes the country of the late Karel Čapek, who not only coined the term "robot" but was also one of the premier thinkers who warned of the unintended consequences of technological development. The best way of dealing with waste is not searching for a miracle solution that would make waste vanish, but finding ways to learn how to live with waste in the best possible way.

The knowledge and experience of those in the lowest hierarchical positions in waste management are useful information sources for the planning of new waste infrastructures as well as their everyday functioning. The contribution of informal waste pickers is unrecognized. Around the world, representatives of associations of informal waste pickers or workers with direct experience in waste infrastructures, for example, are becoming commonplace in policymaking. This is currently hardly thinkable in the Czech context. For example, waste pickers are often seen as criminals and troublemakers who hinder official waste management and cause fires. My recent attempt to introduce a policy paper at a national level has been met with varied reactions. On the one hand, multiple actors appreciated a new perspective on the familiar topic. On the other hand, a normative technoscientific view on waste management and citizens' respect for formal rules have been placed above the need to understand informal workers and mobilize them in ways that can be beneficial for everybody.

Paradoxically, some forms of informal economy would benefit from greater tolerance. This is particularly the case for the various scavengers who increase recycling rates, especially metals, while working at the edge or beyond the legal frameworks of waste infrastructures. In the current complex geopolitical situation, when sources of raw materials such as copper, aluminum, tin, lead, or iron may be challenging to procure, it would make sense to give these actors space to operate rather than tightening the regulatory framework, criminalizing, and punishing them.

It is short-sighted to rely solely on the commercial sector to solve waste-related challenges. These challenges are often long-term. Even the most common municipal solid waste can persist for hundreds of years, affecting future generations. With increasing toxicity, the time horizon is getting even longer. For this reason, we cannot rely entirely on solutions built on the logic of profit within a limited time horizon and externalizing the problem into the future. This is one of the dangers of the contemporary uncritical love affair with the circular economy. As I demonstrated, circular economic models have a strong tendency to postpone the solving of key problems to the future. Following the inspiration associated with fluidity and flows, I join Myra Hird (2021: 225) in her call to look "upstream" rather than continuing to blame consumers situated downstream as usual.

Waste can be thought of similarly as a commitment to care for both tangible and intangible heritage, as well as libraries and individuals who are sick or disabled. The primary imperative is to care for the entity that requires care, and secondarily it is a space where profit might be generated. Instead of outsourcing the problem to multinational corporations, it would be more appropriate to mobilize local forms of commitment to care for the waste generated by local communities. This thinking could also lead to less waste mobility, a more environmentally sensitive approach that aligns with the EU's policy of not externalizing the problem by exporting waste outside the EU.

* * *

It is June 2016, and I visit the Pureland landfill as usual. Lenka, the lady at the gate, informs me that there is nothing new; only the number of incoming trucks is decreasing. I walk toward the tipping zone. Petr is there, but he is in a bad mood. His greeting is cold. A few minutes later, he tells me they learned a few days ago that the landfill would be closed at the end of August. I do not know what to say as I can see how hard it is for him to process the news. Well, it is not the first time the workers have talked about a shutdown, and the decision has been postponed multiple times. However, this time, Petr seems concerned. "They already cut our working hours," he says. I look at him, and something is not right. It takes me a few seconds to realize he got rid of his mullet! Gosh, his legendary haircut is gone. It seems that the situation is critical. When I ask, he explains that it is a consequence of his new job. He took another part-time job as a truck driver at the site, where the ashes from the heating plant are disposed of: "The ash literally flows. You get your head full of it when it gets into the air. So, I had to cut my hair." After all those years at the landfill and careful showering at the end of his shifts to remove the stench from his mullet, he finally made this significant change.

We walk to the edge of the tipping zone. Petr swings his right arm toward the colorful area full of garbage and says, "You can see it right here. Look at that." He says that with disappointment. I can tell that the layer of fresh garbage is thin and has not changed much since I was here a few days ago. We go down to walk through garbage, as usual, to see whether anything interests us. Suddenly, I see large green sheets of metal, then a distorted window, a wheel ... I hold my breath. It is a part of our caravan where we used to sit and chat during rainy days. I completely overlooked that it was not at the edge of the tipping zone as usual. It is hard to resist sadness. Petr does not have to say anything. It is clear what he thinks about. All our jokes, stories, cutting cables, repairing little things, all that is gone. Petr turns to me and says dryly, "Our boss ordered us to destroy the caravan." Apparently, they rolled over it with the front-loader. The caravan finally follows the destiny of all the things that come here. Sooner or later, they end up crushed under the wheels.

The guys carried on working there till the next spring, so the shutdown came later than expected. But this day marked clearly the future inevitability of the shutdown. It was not just the usual complaints and guesses about whether the company would close the landfill or not. The destruction of the caravan changed something. The company could leave it there till the very end, but the manager decided to start the closing process with a radical act of elimination. The caravan had to be crushed and covered with new layers of incoming garbage. Eventually, it disappeared. Really? ...

EPILOGUE

I want to describe some of my drivers and biases in a more personal manner here. I hope readers will appreciate the opportunity to see the book in the context of information that is not immediately apparent.

My family background likely influenced my interest in the lives of people working in industrial environments. My father spent forty-four years in a coking plant as a worker. We lived in Ostrava, the largest city in the north-east, whose vernacular understanding as a "steel heart of the republic" is telling. The steel industry was inextricably linked to hard coal mining. As a child and teenager, I was uninterested in my dad's work. Omnipresent communist exaggeration of workers' importance in the pre-1989 era, the everyday factory smell of my dad's suede jacket and leather bag, as well as his frequent cough, were rather annoying. I recall quarreling about that when I was a rebellious teenager. At the same time, I knew my dad's colleagues and friends. I interacted with them and listened to their stories. I could recognize their habitus immediately from the way the guys moved, greeted each other, and talked, their vague and redundant expressions full of adjectives, respect for physical strength and resistance, or their love of sports. When I began interacting with the workers in the opposite part of the country two decades later, I realized that the environment was surprisingly familiar. I was able to approach them and persuade them to tolerate me despite their prevalent disdain for white-collar people and academics. Perhaps it could be because of my previous experience and ability to switch easily to the workers' way of talking and behaving. Although it was undoubtedly an advantage in terms of finding ways to get close to people, it also pushed me to think about my juvenile experience. There is little doubt that my view of my father's working life influenced my perception of the people who perform manual labor at places of waste disposal. My salvaging tendency to capture the vanishing world of people working at landfills is partially motivated by my internal drive to describe the broader field of workers my family was part of. Readers of this book should bear this in mind.

When reading Andrew Sanchez's epilogue to his book *Criminal Capital* (2016) an essential point about ethnography as an adventure captured my attention. Building upon Contreras's (2013) notion of "Cowboy Ethnographies," Sanchez warns about the researcher's narcissism that fetishizes danger and adventure at the expense of the analytical work. This critical point prompted me to re-evaluate the application of my experience in the book and compelled me to think critically about my motivations. Despite the self-centered focus of the following text, I believe it is fair to explain why someone may fall in love with places where mass waste is disposed of.

I must admit that I am a person who loves danger: speed, heights, and rough environmental conditions. As a child, I competed in downhill skiing. In high school and university, I became immersed in rock climbing and mountaineering. Being in an adrenaline-filled environment of mass disposal was rather a positive aspect of the research for me, especially at the beginning of my research. This changed over time with increasing knowledge of the forms of living associated with landfills. I must reflect on the fundamental difference between my voluntary immersion in such a world and the motivations of those who have been exposed to the sharp teeth of waste streams for years. Neither waste pickers nor landfill workers were forced to work with garbage. Still, they could not simply escape and move to the academic world like me. Their choices were constrained. They depended on waste streams to live their lives and provide resources for their families. Although I mentioned earlier that there was a positive affective dimension of searching for valuable finds in garbage, I do not recall that any of my interlocutors mentioned danger as a positive emotion. They had to accept the danger as an inherent part of their lives, and they would certainly welcome the possibility of reducing it.

Beyond the level of my adrenaline drivers, there was a sense of epistemological commitment to experiencing what those who studied might have experienced. I was deeply touched by Kathleen Millar's ethnography of *catadores* in her book *Reclaiming the Discarded*. Although one might consider her descriptions of extreme exposure to pollution and danger as falling into the "cowboy" kind of ethnography, regardless of gender difference, her texts convey something crucial for me. They reveal what certain humans go through and the obstacles they face in order to live their lives. To understand others, one needs to experience what their practices feel like. Consequently, participation in activities that can be disgusting, exhausting, or dangerous may carry a sense of adventure. Like anything else, it is about balance. I believe that when an ethnographer uses vivid, engaging descriptions to convey a message that fosters a deeper understanding of others' lives, it is justified. This is the approach that I applied to this book.

REFERENCES

Abu Ghosh, Yasar. 2023. "Introduction: Of loss and Silence." In *Romani Chronicles of COVID-19: Testimonies of Harm and Resilience*, ed. Paloma Gay y Blasco and Martin Fotta, 247–58. Oxford: Berghahn.

Abu-Rish, Ziad. 2015. "Garbage Politics." *Middle East Report* 45 (winter 2015). Retrieved 23 March 2016 from www.merip.org/mer/mer277/garbage-politics.

Acedo, Sara S. 2015. "A Way of Life Flowing in the Interstices of Major Economy: Gypsy Horse Dealers in the Alentejo, Portugal." In *Gypsy Economy: Romani Livelihoods and Notions of Worth in the 21st Century*, ed. Micol Brazzabeni, Manuela Ivone Cunha, and Martin Fotta, 68–87. New York: Berghahn.

Alexander, Catherine. 2009. "Illusions of Freedom: Polanyi and Third Sector Recycling Schemes." In *Market and Society: The Great Transformation Today*, ed. Chris Hann and Keith Hart, 221–39. Cambridge: Cambridge University Press.

Alexander, Catherine. 2016. "When Waste Disappears, or More Waste Please!" *RCC Perspectives* 1: 31–40.

Alexander, Catherine. 2023. "A Chronotope of Expansion: Resisting Spatio-temporal Limits in a Kazakh Nuclear Town." *Ethnos* 88(3): 467–90.

Alexander, Catherine, and Patrick O'Hare. 2023. "Waste and Its Disguises: Technologies of (Un)Knowing." *Ethnos* 88(3): 419–43.

Alexander, Catherine, and Joshua Reno. 2012. "Introduction." In *Economies of Recycling: The Global Transformation of Materials, Values and Social Relations*, ed. C. Alexander and J. Reno, 1–32. London: Zed Books.

Alexander, Catherine, and Andrew Sanchez, eds. 2018. *Indeterminacy: Waste, Value, and the Imagination.* New York: Berghahn Books.

Alexander, Kelly. 2024. *Truffles and Trash: Recirculating Food in a Social Welfare State.* Chapel Hill: University of North Carolina Press.

Alibardi, Luca, and Raffaello Cossu. 2018. "6.1 - Leachate Generation Modeling." In *Solid Waste Landfilling*, ed. Raffaello Cossu and Rainer Stegmann, 229–45. Elsevier.

Alves, Bruna. 2023. "Waste Management Market Value Worldwide 2022–2030." Retrieved 9 June 2023 from www.statista.com/statistics/246178/projected-global-waste-management-market-size/.

Amit, Vered. 2000. "Introduction: Constructing the field." In *Constructing the Field*, ed. Vered Amit, 1–18. London, Routledge.

Antošová, Veronika, Lucie Veselá, Andrea Králiková, and Lenka Procházková. 2024. "Food Waste in Generation Z: The Impact of Social Media and the Ineffectiveness of Current Promotional Strategies." *Waste Forum* (4): 274–90.

Appadurai, Arjun. 1986. "Introduction: Commodities and the Politics of Value." In *The Social Life of Things: Commodities in Cultural Perspective*, ed. Arjun Appadurai, 3–63. Cambridge: Cambridge University Press.

Balaš, Nikola. 2018. "Academia Without Contention? The Legacy of Czechoslovak Ethnography and Folklore Studies in Czech Anthropology." *Czech Sociological Review* 54(3): 343–70.

Balaš, Nikola. 2024. *An Ethnographic Chiefdom: Epistemic Arrest and Knowledge Production in Czechoslovak Ethnography (1969–1989)*. New York: Berghahn Books.

Ballinger, Ann, William Shanks, Dominic Hogg, Chris Sherrington, and Lauren Duffield. 2020. *Greenhouse Gas and Air Quality Impacts of Incineration and Landfill*. Report for ClientEarth. Bristol: Eunomia.

Barry, Andrew. 2001. *Political Machines: Governing a Technological Society*. London: Athlone.

Barry, Andrew. 2006. "Technological Zones." *European Journal of Social Theory* 9(2): 239–53.

Basso, Keith H. 1996. *Wisdom Sits in Places: Landscape and Language Among the Western Apache*. Abuquerque: UNM Press.

Bateson, Gregory. 1936. *Naven*. Cambridge: Cambridge University Press.

Bateson, Gregory. 2000. *Steps to an Ecology of Mind*. Chicago: University of Chicago Press.

Bauman, Zygmunt. 2000. *Liquid Modernity*. Cambridge: Polity.

Bauman, Zygmunt. 2004. *Wasted Lives: Modernity and its Outcasts*. Malden: Polity.

Baviskar, Amita, and Vinay Gidwani. 2019. "The Lives of Waste and Pollution." *Economic & Political Weekly* 54. Retrieved 20 December 2019 from www.epw.in/journal/2019/47/review-urban-affairs/lives-waste-and-pollution.html.

Bear, Laura. 2013. "The Antinomies of Audit: Opacity, Instability and Charisma in the Economic Governance of a Hooghly Shipyard." *Economy and Society* 42(3): 375–97.

Beaven, Richard P., and Keith Knox. 2018. "Chapter 12.2—Leachate Recirculation: Design, Operation, and Control." In *Solid Waste Landfilling*, ed. Raffaello Cossu and Rainer Stegmann, 703–27. Elsevier.

Benešová, Libuše, Bohumil Černík, Petra Hnaťuková, Zdenka Kotoulová, and Martina Vrbová. 2008. "Výzkum vlastností komunálních odpadů a optimalizace jejich využití." *Zpráva pro závěrečný kontrolní den v roce 2008 VaV SP/2f1/132/08*. Praha: Přírodovědecká fakulta, Univerzita Karlova.

Berdahl, Daphne, and Matti Bunzl. 2010. *On the Social life of Postsocialism: Memory, Consumption, Germany*. Bloomington: Indiana University Press.

Berg, Anne. 2023. "A Rubbished World: White Supremacy's Complicated Love Affair with Garbage." *Journal of Genocide Research* 25(1): 46–61.

Berliner, David. 2022. *Devenir autre: Hétérogénéité et plasticité du soi*. Paris: La Découverte.

Berliner, David, Michael Lambek, Richard Shweder, Richard Irvine, and Albert Piette. 2016. "Anthropology and the Study of Contradictions." *HAU: Journal of Ethnographic Theory* 6(1): 1–27.

Bernard, Josef, and Petr Vašát. 2015. "Forming Communities or Social Integration? A Personal Network Analysis of Ukrainian Immigrants in Pilsen." *Czech Sociological Review* 51(2): 199–225.

Bille, Mikkel, Frida Hastrup, and Tim Flohr Sørensen, eds. 2010. *An Anthropology of Absence: Materializations of Transcendence and Loss.* New York: Springer.

Billig, Michael. 2019. *Schwarz. Rot. Müll: Die schmutzigen Deals der deutschen Müllmafia.* Freiburg: Verlag Herder GmbH.

Bilska, Beata, Marzena Tomaszewska, and Danuta Kolozyn-Krajewska. 2024. "Food Waste in Polish Households: Characteristics and Sociodemographic Determinants on the Phenomenon. Nationwide Research." *Waste Management* 176: 30–40.

Bize, Amiel. 2020. "The Right to the Remainder: Gleaning in the Fuel Economies of East Africa's Northern Corridor." *Cultural Anthropology* 35(3): 462–86.

Bloch, Maurice. 1998. *How We Think They Think: Anthropological Approaches to Cognition, Memory, and Literacy.* Boulder: Westview Press.

Bloch, Maurice, and Jonathan Parry. 1982. "Introduction: Death and the Regeneration of Life." In *Death and the Regeneration of Life*, ed. Maurice Bloch and Jonathan Parry, 1–44. Cambridge: Cambridge University Press.

Block, Fred L., and Margaret R. Somers. 2014. *The Power of Market Fundamentalism: Karl Polanyi's Critique.* Cambridge: Harvard University Press.

Bovensiepen, Judith, and Mathijs Pelkmans. 2020. "Dynamics of Wilful Blindness: An Introduction." *Critique of Anthropology* 40(4): 387–402.

Bowker, Geoffrey C., and Susan Leigh Star. 1999. *Sorting Things Out: Classification and its Consequences.* Cambridge: MIT Press.

Brazzabeni, Micol, Manuela Ivone Pereira da Cunha, and Martin Fotta, eds. 2015a. *Gypsy Economy: Romani Livelihoods and Notions of Worth in the 21st Century.* New York: Berghahn Books.

Brazzabeni, Micol, Manuela Ivone Pereira da Cunha, and Martin Fotta. 2015b. "Introduction: Gypsy Economy." In *Gypsy Economy: Romani Livelihoods and Notions of Worth in the 21st Century*, ed. Micol Brazzabeni, Manuela Ivone Pereira da Cunha, and Martin Fotta, 1–30. New York: Berghahn Books.

Breman, Jan. 1976. "A Dualistic Labour System? A Critique of the 'Informal Sector' Concept: I: The Informal Sector." *Economic and Political Weekly* 11(48): 1870–76.

Bridger, Susan, and Frances Pine, eds. 1998. *Surviving Post-Socialism: Local Strategies and Regional Responses in Eastern Europe and the Former Soviet Union.* London: Routledge.

Brković, Čarna. 2015. "Management of Ambiguity: Favours and Flexibility in Bosnia and Herzegovina." *Social Anthropology* 23(3): 268–82.

Brković, Čarna, and Andrew Hodges. 2015. "Rethinking World Anthropologies Through Fieldwork: Perspectives on 'Extended Stay' and 'Back-and-Forth' Methodologies." *Anthropological Notebooks* 21(1): 107–20.

Brož, Luděk. 2024. *Evil Spirits and Rocket Debris: In Search of Lost Souls in Siberia.* New York: Berghahn.

Brunclíková, Lenka. 2017a. "Recyclable Waste as a Marker of Everyday Life Routines." In *Archaeologies of Waste: Encounters with the Unwanted*, ed. Daniel Sosna and Lenka Brunclíková, 100–17. Oxford: Oxbow Books.

Brunclíková, Lenka. 2017b. "Waste and Value in the Post-Socialist Space." *Anthropologie (Brno)* 55(3): 385–98.

Bryant, Rebecca, and Daniel M. Knight. 2019. *The Anthropology of the Future.* Cambridge: Cambridge University Press.

Bubandt, Nils, and Rane Willerslev. 2015. "The Dark Side of Empathy: Mimesis, Deception, and the Magic of Alterity." *Comparative Studies in Society and History* 57(1): 5–34.

Buchowski, Michal. 2004. "Hierarchies of Knowledge in Central-Eastern European Anthropology." *Anthropology of East Europe Review* 22(2): 5–14.

Buchowski, Michal. 2006. "The Specter of Orientalism in Europe: From Exotic Other to Stigmatized Brother." *Anthropological Quarterly* 79(3): 463–82.

Bugnyar, Thomas, and Kurt Kotrschal. 2002. "Observational Learning and the Raiding of Food Caches in Ravens, Corvus Corax: Is it 'Tactical' Deception?" *Animal Behaviour* 64(2): 185–95.

Burawoy, Michael, and Katherine Verdery, eds. 2000. *Uncertain Transition: Ethnographies of Change in the Postsocialist World.* Lanham: Rowman & Littlefield Publishers.

Butt, Waqas H. 2020a. "Accessing Value in Lahore's Waste Infrastructures." *Ethnos* 88: 533–53.

Butt, Waqas H. 2020b. "Waste Intimacies." *American Ethnologist* 47: 234–48.

Butt, Waqas H. 2023. *Life Beyond Waste: Work and Infrastructure in Urban Pakistan.* Stanford: Stanford University Press.

Calabro, P. S., E. Gentili, C. Meoni, S. Orsi, and D. Komilis. 2018. "Effect of the Recirculation of a Reverse Osmosis Concentrate on Leachate Generation: A Case Study in an Italian Landfill." *Waste Management* 76: 643–51.

Caldwell, Melissa L. 2009. "Introduction: Food and Everyday Life After State Socialism." In *Food and Everyday Life in the Postsocialist World*, ed. Melissa L. Caldwell, 1–28. Bloomington: Indiana University Press.

Carenzo, Sebastián, and Lucas Becerra. 2024. "Disruptive but Normalizing?: What the Formalization of Informality Can Tell Us About the Circular Economy in the Global South." In *Circular Economies in an Unequal World: Waste, Renewal and the Effects of Global Circularity*, ed. Patrick O'Hare and Dagna Rams, 113–32. London: Bloomsbury.

Carrier, James G. 2017. "Moral Economy: What's in a Name." *Anthropological Theory* 18(1): 18–35.

Carter, Francis W. 1985. "Pollution Problems in Post-War Czechoslovakia." *Transactions of the Institute of British Geographers* 10(1): 17–44.

Castaneda, Quetzil E. 2006. "Ethnography in the Forest: An Analysis of Ethics in the Morals of Anthropology." *Cultural Anthropology* 21(1): 121–45.

Cervinkova, Hana, Michal Buchowski, and Zdeněk Uherek, eds. 2015. *Rethinking Ethnography in Central Europe.* New York: Palgrave Macmillan.

Chelcea, Liviu. 2015. "Postindustrial Ecologies: Industrial Rubble, Nature and the Limits of Representation." *Parcours Anthropologiques* 10: 186–201.

Cherkaev, Xenia A. 2023. *Gleaning for Communism: The Soviet Socialist Household in Theory and Practice.* Ithaca: Cornell University Press.

Cohen, Henri, and Claire Lefebvre. 2005. "Bridging the Category Divide: Introduction to the First Edition." In *Handbook of Categorization in Cognitive Science*, ed. Henri Cohen and Claire Lefebvre, 1–15. Amsterdam: Elsevier.

Cohen, Stanley. 1972. *Folk Devils and Moral Panics: The Creation of the Mods and Rockers, Sociology and the Modern World.* London: MacGibbon and Kee.

Contreras, Randol. 2013. *The Stickup Kids: Race, Drugs, Violence, and the American Dream.* Berkeley: University of California Press.

Corvellec, Hervé, ed. 2025. *Waste as a Critique*. New York: Oxford University Press.

Corvellec, Hervé, and Johan Hultman. 2012. "From 'Less Landfilling' to 'Wasting Less'." *Journal of Organizational Change Management* 25(2): 297–314.

Corvellec, Hervé, Alison F. Stowell, and Nils Johansson. 2022. "Critiques of the Circular Economy." *Journal of Industrial Ecology* 26: 421–32.

Cossu, Raffaello. 2018. "2.2 - Mass Balance of Contaminants: A Key for Modern Landfill Design." In *Solid Waste Landfilling*, ed. Raffaello Cossu and Rainer Stegmann, 73–88. Amsterdam: Elsevier.

Crumley, Carole L. 1995. "Heterarchy and the Analysis of Complex Societies." In *Heterarchy and the Analysis of Complex Societies*, ed. Robert M. Ehrenreich, Carole L. Crumley and Janet E. Levy, 1–5. Arlington: American Anthropological Association.

CSO (Czech Statistical Office). 2016. "Average Wages—3. Quarter of 2016." Retrieved 20 August 2024 from https://csu.gov.cz/rychle-informace/average-wages-3-quarter-of-2016.

Daněk, Petr, and Petr Jehlička. 2020. "Quietly Degrowing: Food Self-Provisioning in Central Europe." In *Food for Degrowth*, ed. Petr Daněk and Petr Jehlička, 33–44. London: Routledge.

Das, Veena. 2015. "What Does Ordinary Ethics Look Like." *Four Lectures on Ethics: Anthropological Perspectives*, 53–125. Chicago: HAU Books.

De L'Estoile, Benoît. 2014. "Money Is Good, but a Friend Is Better." *Current Anthropology* 55(S9): S62–S73.

De León, Jason. 2013. "Undocumented Migration, Use Wear, and the Materiality of Habitual Suffering in the Sonoran Desert." *Journal of Material Culture* 18(4): 321–45.

De León, Jason. 2015. *The Land of Open Graves: Living and Dying on the Migrant Trail.* Oakland: University of California Press.

Deacon, Terrence W. 2012. *Incomplete Nature: How Mind Emerged from Matter.* New York: W.W. Norton & Co.

Delanová, Šárka. 2015. "Nepředvídatelnost, flexibilita a umírněná tekutost v globální továrně." *Cargo Journal* 13(1–2): 95–115.

Deleuze, Gilles, and Félix Guattari. 1987. *A Thousand Plateaus: Capitalism and Schizophrenia.* Minneapolis: University of Minnesota Press.

Dezhamkhooy, Maryam. 2024. "Wandering Waste: Informal Waste Work and the Externalization of People and Landscapes in Iran." *Current Anthropology* 65(5): 856–75.

Dikovic, Jovana. 2016. "Gleaning: Old Name, New Practice." *The Journal of Legal Pluralism and Unofficial Law* 48(2): 302–21.

Doherty, Jacob. 2019. "Filthy Flourishing." *Current Anthropology* 60(S20): S321–S332.

Doherty, Jacob. 2022. *Waste Worlds: Inhabiting Kampala's Infrastructures of Disposability.* Oakland: University of California Press.

Dohotaru, Aadrian O. 2013. "Performative Anthropology. The Case of the Pata-Rât Ghetto." *Studia Universitatis Babes-Bolyai-Sociologia* 58(2): 193–215.

Douglas, Mary. 2002[1966]. *Purity and Danger: An Analysis of the Concepts of Pollution and Taboo.* London: Routledge.

Drahokoupil, Jan. 2009. *Globalization and the State in Central and Eastern Europe: the Politics of Foreign Direct Investment.* London: Routledge.

Dunajeva, Jekatyerina, and Joanna Kostka. 2021. "Racialized Politics of Garbage: Waste Management in Urban Roma Settlements in Eastern Europe." *Ethnic and Racial Studies* 45(1): 90–112.

Dunn, Elizabeth C. 2004. *Privatizing Poland: Baby Food, Big Business, and the Remaking of Labor.* Ithaca: Cornell University Press.

Dzenovska, Dace 2018. *School of Europeanness: Tolerance and Other Lessons in Political Liberalism in Latvia.* Ithaca: Cornell University Press.

Edwards, David B. 2017. *Caravan of Martyrs: Sacrifice and Suicide Bombing in Afghanistan.* Oakland: University of California Press.

EEA (European Environment Agency). 2021. *Knowledge for Action: Empowering the Transition to a Sustainable Europe, EEA Report No 10/2021.* Luxembourg: European Environment Agency.

EEC (Council of the European Economic Community). 1975. *Council Directive 75/442/ EEC of 15 July 1975 on waste.* Retrieved 10 June 2024 from https://eur-lex.europa.eu/eli/dir/1975/442/oj/eng.

Egüez, Alejandro. 2021. "Compliance with the EU Waste Hierarchy: A Matter of Stringency, Enforcement, and Time." *Journal of Environmental Management* 280: 111672.

Ehrig, Hans-Jürgen, and Rainer Stegmann. 2018. "Chapter 10.2 - Leachate Quality." In *Solid Waste Landfilling,* ed. Raffaello Cossu and Rainer Stegmann, 511–39. Amsterdam: Elsevier.

Eiselt, Horts A. 2007. "Locating Landfills—Optimization vs. Reality." *European Journal of Operational Research* 179(3): 1040–49.

Ellen, Roy. 1988. Fetishism. *Man* 23(2): 213–35.

Enviweb. 2004. "Skládka je časovanou bombou." Retrieved 14 April 2022 from www.enviweb.cz/48926.

Eriksen, Thomas Hylland. 2016. *Overheating: an Anthropology of Accelerated Change.* London: Pluto Press.

Escobar, Arturo. 1999. "After Nature: Steps to an Antiessentialist Political Ecology." *Current Anthropology* 40(1): 1–30.

EU (European Union). 2013. "2013/727/EU: Establishing a Format for Notifying the Information on the Adoption and Substantial Revisions of the Waste Management Plans and the Waste Prevention Programmes." *Official Journal of the European Union*: L329/44–48.

EU (European Union). 2018a. *Council Directive 1999/31/EC of 26 April 1999 on the Landfill of Waste, 1999/31/EC.* Brussels: European Union.

EU (European Union). 2018b. *Directive 2008/98/EC of the European Parliament and of the Council of 19 November 2008 on Waste and Repealing Certain Directives, 2008/98/EC.* Brussels: European Union.

Evans-Pritchard, Edward E. 1937. *Witchcraft, Oracles and Magic among the Azande.* Oxford: The Clarendon Press.

Fabian, Johannes. 2006. "The Other Revisited: Critical Afterthoughts." *Anthropological Theory* 6(6): 139–52.

Fahmi, Wael, and Keith Sutton. 2010. "Cairo's Contested Garbage: Sustainable Solid Waste Management and the Zabaleen's Right to the City." *Sustainability* 2(6): 1765–83.

Fair, Hannah, Viola Schreer, Paul Keil, Laur Kiik, and Niki Rust. 2023. "Dodo Dilemmas: Conflicting Ethical Loyalties in Conservation Social Science Research." *Area* 55(2): 245–53.

Feyerabend, Paul. 1975. *Against Method: Outline of an Anarchistic Theory of Knowledge.* Atlantic Highlands: Humanities Press.

Filčák, Richard, and Ondřej Ficeri. 2021. "Making the Ghetto at Luník IX in Slovakia: People, Landfill, and the Myth of the Urban Green Space." *Sociologický časopis/Czech Sociological Review* 57(3): 1–21.

Fine, Michelle. 1994. "Working the Hyphens: Reinventing Self and Other in Qualitative Research." In *Handbook of Qualitative Research*, ed. Norman K. Denzin and Yvonna S. Lincoln, 70–82. Thousand Oaks: Sage.

Fishberg, Rachel, Anton Grau Larsen, and Kristoffer Kropp. 2023. "The 'Where' of EU Social Science Collaborations: How Epistemic Inequalities and Geopolitical Power Asymmetries Persist in Research About Europe." *The Sociological Review* 72(5): 975–97.

Fotta, Martin. 2019. "'Only the Dead Don't Make the Future': Calon Lives Between Non-Gypsies and Death." *Journal of the Royal Anthropological Institute* 25(3): 587–605.

Franc, Pavel, and Jiří Nezhyba. 2007. *Zahraniční investice a CzechInvest jako faktory destabilizující demokratický právní stát.* Brno: Ekologický právní servis.

Fredericks, Rosalind. 2018. *Garbage Citizenship: Vital Infrastructures of Labor in Dakar, Senegal.* Durham: Duke University Press.

Frederiksen, Martin D. 2015. "The Last Honest Bandit: Transparency and Spectres of Illegality in the Republic of Georgia." In *Ethnographies of Grey Zones in Eastern Europe: Relations, Borders and Invisibilities*, ed. Ida H. Knudsen and Martin D. Frederiksen, 157–70. London: Anthem Press.

Frederiksen, Martin D., and Ida H. Knudsen. 2015. "Introduction: What is a Grey Zone and Why is Eastern Europe One." In *Ethnographies of Grey Zones in Eastern Europe: Relations, Borders and Invisibilities*, ed. Ida H. Knudsen and Martin D. Frederiksen, 1–22. London: Anthem Press.

Furniss, Jamie. 2017. "What Type of Problem is Waste in Egypt?" *Social Anthropology* 25(3): 301–17.

Gan, Elaine, Anna Tsing, Heather Anne Swanson, and Nils Ole Bubandt. 2017. "Introduction: Haunted landscapes of the Anthropocene." In *Arts of Living on a Damaged Planet*, ed. Anna L. Tsing, Heather A. Swanson, Elaine Gan, and and Nils O. Bubandt, G1–G14. Minneapolis: University of Minnesota Press.

Garcier, Romain. 2012. "One Cycle to Bind Them All? Geographies of Nuclearity in the Uranium Fuel Cycle." In *Economies of Recycling: Global Transformations of Materials, Values and Social Relations*, ed. Catherine Alexander and Joshua Reno, 76–97. London: Zed Books.

Geissdoerfer, Martin, Paulo Savaget, Nancy M. P. Bocken, and Erik Jan Hultink. 2017. "The Circular Economy: A New Sustainability Paradigm?" *Journal of Cleaner Production* 143: 757–68.

Gidwani, Vinay. 2015. "The Work of Waste: Inside India's Infra-Economy." *Transactions of the Institute of British Geographers* 40(5): 575–95.

Gille, Zsuzsa. 2007. *From the Cult of Waste to the Trash Heap of History: The Politics of Waste in Socialist and postsocialist Hungary.* Bloomington: Indiana University Press.

Gille, Zsuzsa. 2012. "From Risk to Waste: Global Food Waste Regimes." *The Sociological Review* 60(2_suppl): 27–46.

Gille, Zsuzsa. 2013. "Is There an Emancipatory Ontology of Matter? A Response to Myra Hird." *Social Epistemology Review and Reply Collective* 2(4): 1–6.

Gille, Zsuzsa. 2016. *Paprika, Foie Gras, and Red Mud: The Politics of Materiality in the European Union.* Bloomington: Indiana University Press.

Gille, Zsuzsa. 2025. "The Socialocene: From Capitalocene to Transnational Waste Regimes." *Antipode* 57(5): 1705–24.

Gillespie, Susan D. 2001. "Personhood, Agency, and Mortuary Ritual: A Case Study from the Ancient Maya." *Journal of Anthropological Archaeology* 20(1): 73–112.

Girard, René. 1986[1982]. *The Scapegoat.* Baltimore: Johns Hopkins University Press.

Gmelch, Sharon B. 1986. "Groups that Don't Want in: Gypsies and Other Artisan, Trader, and Entertainer Minorities." *Annual Review of Anthropology* 15: 307–30.

Gödel, Kurt. 1992[1931]. *On Formally Undecidable Propositions of Principia Mathematica and Related Systems.* New York: Dover Publications.

Godelier, Maurice. 1999. *The Enigma of the Gift.* Chicago: University of Chicago Press.

González-Ruibal, Alfredo. 2008. "Time to Destroy: An Archaeology of Supermodernity." *Current Anthropology* 49(2): 247–79.

Gordillo, Gastón. 2014. *Rubble: the Afterlife of Destruction.* Durham: Duke University Press.

Graeber, David. 2007. *Possibilities: Essays on Hierarchy, Rebellion, and Desire.* Oakland: AK Press.

Graeber, David. 2012. "Afterword: The Apocalypse of Objects: Degradation, Redemption, and Transcendence in the World of Consumer Goods." In *Economies of Recycling: The Global Transformation of Materials, Values and Social Relations*, ed. Catherine Alexander and Joshue Reno, 277–90. London: Zed Books.

Gregson, Nicky, and Mike Crang. 2016. "Illicit Economies: Customary Illegality, Moral Economies and Circulation." *Transactions of the Institute of British Geographers* 42(2): 206–19.

Gregson, Nicky, Mike Crang, Sara Fuller, and Helen Holmes. 2015. "Interrogating the Circular Economy: the Moral Economy of Resource Recovery in the EU." *Economy and Society* 44(2): 218–43.

Grindal, Bruce. 2011. "Confrontation, Understanding, and Friendship in a Redneck Culture." *Anthropology and Humanism* 36(1): 89–100.

Grygar, Jakub. 2016. *Děvušky a cigarety: o hranicích, migraci a moci.* Praha: SLON.

Gudeman, Stephen. 2008. *Economy's Tension: the Dialectics of Community and Market.* New York: Berghahn.

Gudeman, Stephen. 2016. *Anthropology and Economy.* Cambridge: Cambridge University Press.

Gudeman, Stephen, and Chris Hann, eds. 2015. *Oikos and Market: Explorations in Self-Sufficiency after Socialism.* New York: Berghahn.

Günel, Gökçe, Saiba Varma, and Chika Watanabe. 2020. "A Manifesto for Patchwork Ethnography." *Member Voices, Fieldsights* June 9. Retrieved 8 October 2022 from https://culanth.org/fieldsights/a-manifesto-for-patchwork-ethnography.

Günel, Gökçe, and Chika Watanabe. 2024. "Patchwork Ethnography." *American Ethnologist* 51(1): 131–39.

Gutberlet, Jutta, Santiago Sorroche, Angela Martins Baeder, Patrik Zapata, and María José Zapata Campos. 2021. "Waste Pickers and Their Practices of Insurgency and Environmental Stewardship." *The Journal of Environment & Development* 30(4): 369–94.

Guyer, Jane I. 2004. *Marginal Gains: Monetary Transactions in Atlantic Africa.* Chicago: University of Chicago Press.

Hacking, Ian. 1990. *The Taming of Chance.* Cambridge: Cambridge University Press.

Hage, Ghassan. 2021. "Introduction: States of Decay." In *Decay*, ed. Ghassan Hage, 1–15. Durham: Duke University Press.

Hage, Ghassan. 2023. "The Anthropological and the Consequential." *American Anthropologist* 125(1): 168–70.

Hájek, Martin, Petr Kaderka, and Jiří Nekvapil. 2019. *Kdo šetří, má za tři?: Diskurz šetrnosti v proměnách české společnosti*. Praha: SLON.

Hall, Stuart, Chase Critcher, Tony Jefferson, John Clarke, and Brian Roberts. 1978. *Policing the Crisis: Mugging, Law and Order and the State*. London: Macmillan.

Haluzík, Radan, ed. 2020. *Město naruby: vágní terén, vnitřní periferie a místa mezi místy*. Praha: Academia.

Hann, Chris, ed. 2002. *Postsocialism: Ideals, Ideologies, and Practices in Eurasia*. London: Routledge.

Hann, Chris. 2014. "Why Post-imperial Trumps Post-socialist." In *Anthropology and Nostalgia*, ed. Olivia Angé and David Berliner, 96–122. New York: Berghahn Books.

Hann, Chris. 2016. "A Concept of Eurasia." *Current Anthropology* 57(1): 1–27.

Hann, Chris, Caroline Humphrey, and Katherine Verdery. 2002. "Introduction: Postsocialism as a Topic of Anthropological Investigation." In *Postsocialism: Ideals, Ideologies, and Practices in Eurasia*, ed. Chris Hann, 1–28. London: Routledge.

Harris, Marvin. 2001[1984]. "The Abominable Pig." In *Ethnic and Regional Foodways in the United States: The Performance of Group Identity*, ed. Linda K. Brown and Kay Mussell, 67–79. Knoxville: The University of Tennessee Press.

Hart, Keith. 1973. "Informal Income Opportunities and Urban Employment in Ghana." *The Journal of Modern African Studies* 11(1): 61–89.

Hart, Keith. 2010. "Informal Economy." In *The Human Economy: A Citizen's Guide*, ed. Keith Hart, Jean-Louis Laville and Antonio D. Cattani, 142–53. Cambridge: Polity.

Hartman, Tod. 2008. "States, Markets, and Other Unexceptional Communities: Informal Romanian Labour in a Spanish Agricultural Zone." *Journal of the Royal Anthropological Institute* 14: 496–514.

Harvey, David. 2007. *A Brief History of Neoliberalism*. Oxford: Oxford University Press.

Harvey, David. 2010. *A Companion to Marx's Capital*. London: Verso.

Hastrup, Kirsten. 2004. "Getting it Right: Knowledge and Evidence in Anthropology." *Anthropological Theory* 4(4): 455–72.

Haukanes, Haldis, and Frances Pine. 2021. "Silences and Secrets of Family, Community and the State." *Ethnologia Polona* 42: 99–112.

Hecht, Gabrielle. 2014. *Being Nuclear: Africans and the Global Uranium Trade*. Boston: MIT Press.

Hecht, Gabrielle. 2018. "Interscalar Vehicles for an African Anthropocene: On Waste, Temporality, and Violence." *Cultural Anthropology* 33(1): 109–41.

Hejnal, Ondřej. 2013. "Hilton jako 'fekální dvůr'. Socioprostorové aspekty bezdomovectví." *Sociologický časopis* 49(2): 241–67.

Hendrych, Jiří, Radka Hejralová, Jiří Kroužek, Pavel Špaček, and Jiří Sobek. 2019. "Stabilisation/Solidification of Landfill Leachate Concentrate and its Residue Obtained by Partial Evaporation." *Waste Management* 95: 560–68.

Henig, David. 2012. "Iron in the Soil: Living with Military Waste in Bosnia-Herzegovina." *Anthropology Today* 28(1): 21–23.

Henig, David. 2019. "Living on the Frontline: Indeterminacy, Value, and Military Waste in Postwar Bosnia-Herzegovina." *Anthropological Quarterly* 92(1): 85–110.

Henig, David. 2020. "Emptiness and Deadly Environments. Theorizing the Contemporary, Fieldsites." Editor's Forum in *Cultural Anthropology*. Retrieved 17 December 2020 from https://culanth.org/fieldsights/emptiness-and-deadly-environments.

Henig, David. 2024. "When War Percolates: On Topologies of Earthly Violence in a Planetary Age. In *Porous Becomings: Anthropological Engagements with Michel Serres*, ed. Andreas Bandak and Daniel M. Knight, 135–52. Durham: Duke University Press.

Henig, David and Daniel M. Knight. 2023. "Polycrisis: Prompts for an Emerging Worldview." *Anthropology Today* 39: 3–6.

Henig, David, and Nicolette Makovicky, eds. 2017. Economies of Favour After Socialism. Oxford: Oxford University Press.

Henig, David, and Anna Strhan. 2022. "Introduction. The Good between Social Theory and Philosophy." In *Where is the Good in the World? Ethical Life between Social Theory and Philosophy*, ed. David Henig, Anna Strhan, and Joel Robbins, 1–32. New York: Berghahn.

Hertz, Robert. 1960[1907]. *Death and the Right Hand.* Aberdeen: Cohen & West.

Herzfeld, Michael. 2001. *Anthropology: Theoretical Practice in Culture and Society.* Malden: Blackwell Publishers.

Hetherington, Kevin. 2004. "Secondhandedness: Consumption, Disposal, and Absent Presence." *Environment and Planning D: Society and Space* 22(1): 157–73.

Hird, Myra J. 2012. "Knowing Waste: Towards an Inhuman Epistemology." *Social Epistemology* 26(3–4): 453–69.

Hird, Myra J. 2021. *Canada's Waste Flows.* Montreal: McGill-Queen's University Press.

Hird, Myra J. 2022. *A Public Sociology of Waste.* Bristol: Bristol University Press.

Hirsch, Eric. 1995. "Introduction: Landscape: Between Place and Space." In *The Anthropology of Landscape: Perspectives on Place and Space*, ed. Eric Hirsch and Michael O'Hanlon, 1–30. Oxford: Clarendon Press.

Hoag, Colin, Filippo Bertoni, and Nils Bubandt. 2018. "Wasteland Ecologies: Undomestication and Multispecies Gains on an Anthropocene Dumping Ground." *Journal of Ethnobiology* 38(1): 88–104.

Holmström, Mark. 1984. *Industry and Inequality: The Social Anthropology of Indian Labour.* Cambridge: Cambridge University Press.

Holy, Ladislav. 2003. "The End of Socialism in Czechoslovakia." In *Socialism: Ideals, Ideologies, and Local Practice*, ed. Chris Hann, 252–67. London: Routledge.

Horáková, Pavla. 2018. *Teorie podivnosti* [The Theory of Strangeness]. Praha: Argo.

Hornborg, Alf. 2016. *Global Magic: Technologies of Appropriation from Ancient Rome to Wall Street.* New York: Palgrave.

Hrabal, Bohumil. 1994. *Hlučná samota* [Loud Solitude]. Sebrané spisy Bohumila Hrabala 9, ed. Milan Jankovič. Praha: Pražská imaginace.

Hultman, Johan, Hervé Corvellec, Anne Jerneck, Susanne Arvidsson, Johan Ekroos, Clara Gustafsson, Fay Lundh Nilsson, and Niklas Wahlberg. 2021. "A Resourcification Manifesto: Understanding the Social Process of Resources Becoming Resources." *Research Policy* 50(9): 104297.

Humphrey, Caroline. 2002. *The Unmaking of Soviet Life: Everyday Economies After Socialism.* Ithaca: Cornell University Press.

Humphrey, Caroline. 2012. "Favors and 'Normal Heroes': The Case of Postsocialist Higher Education." *HAU: Journal of Ethnographic Theory* 2(2): 22–41.

Ingold, Tim. 2000. *The Perception of the Environment: Essays on Livelihood, Dwelling and Skill.* London: Routledge.

Ingold, Tim. 2013. "Prospect." In *Biosocial Becomings: Integrating Social and Biological Anthropology*, ed. Tim Ingold and Gisli Palsson, 1–21. Cambridge: Cambridge University Press.

Ingold, Tim, and Gisli Palsson. 2013. *Biosocial Becomings: Integrating Social and Biological Anthropology.* Cambridge: Cambridge University Press.

Ingram, Toni. 2024. "Sexuality-Assemblages, Hyphens, and the In-Between." *Qualitative Inquiry* 31(3–4): 365–72.

Jackson, Anthony. 1987a. *Anthropology at Home.* London: Tavistock Publications.

Jackson, Anthony. 1987b. "Reflections on Ethnography at Home and the ASA." In *Anthropology at Home*, ed. Anthony Jackson, 1–15. London: Tavistock.

Jackson, Deborah D. 2011. "Scents of Place: The Dysplacement of a First Nations Community in Canada." *American Anthropologist* 113(4): 606–18.

Jamieson, William. 2020. "For Granular Geography." *Dialogues in Human Geography* 11(2): 275–93.

Jehlička, Petr. 2002. "Environmental Implications of Eastern Enlargement of the EU: The End of Progressive Environmental Policy?" *RSC No. 2002/23.* San Domenico: European University Institute.

Jehlička, Petr. 2021. "Eastern Europe and the Geography of Knowledge Production: The Case of the Invisible Gardener." *Progress in Human Geography* 45(5): 1218–36.

Jehlička, Petr, Miķelis Grīviņš, Oane Visser, and Bálint Balázs. 2020. "Thinking Food Like an East European: A Critical Reflection on the Framing of Food Systems." *Journal of Rural Studies* 76(May 2020): 286–95.

Jehlička, Petr, Tomáš Kostelecký, and Joe Smith. 2013. "Food Self-Provisioning in Czechia: Beyond Coping Strategy of the Poor: A Response to Alber and Kohler's 'Informal Food Production in the Enlarged European Union' (2008)." *Social Indicators Research* 111(1): 219–34.

Jehlička, Petr, Lucie Veselá, and Lea Kubíčková. 2025. "Where and When Matter: Uncovering the Hidden Influences on Household Food Waste Generation Rates." *British Food Journal* 127(13): 412–25.

Jovanović, Deana. 2021. "Infrastructural Stripping and 'Recycling' of Copper: Producing the State in an Industrial Town in Serbia." *Social Anthropology* 29(3): 586–601.

Juskus, Ryan. 2023. "Sacrifice Zones." *Environmental Humanities* 15(1): 3–24.

Kacperczyk, Anna. 2021. "Phenomenology of Trash." In *Deathworlds to Lifeworlds: Collaboration with Strangers for Personal, Social and Ecological Transformation*, ed. Valerie Malhotra Bentz and James Marlatt, 63–91. Berlin: De Gruyter.

Kaiserová, Ida, and Jiří Mertl. 2016. *Plzeň – město (politické) kultury?Analýza tří případů subpolitického rozhodování v Plzni.* Plzeň.

Kawa, Nicholas C. 2016. What Happens When we Flush? *Anthropology Now* 8(2): 34–43.

Keane, Webb. 2003. "Semiotics and the Social Analysis of Material Things." *Language & Communication* 23(3): 409–25.

Keane, Webb. 2015. *Ethical Life: Its Natural and Social Histories.* Princeton: Princeton University Press.

Keil, Paul G. 2021. "Rank Atmospheres: The More-than-Human Scentspace and Aesthetic of a Pigdogging Hunt." *The Australian Journal of Anthropology* 32(S1): 96–113.

Kideckel, David A. 2008. *Getting by in Postsocialist Romania: Labor, the Body, & Working-Class Culture.* Bloomington: Indiana University Press.

Kim, Eleana J. 2016. "Toward an Anthropology of Landmines: Rogue Infrastructure and Military Waste in the Korean DMZ." *Cultural Anthropology* 31(2): 162–87.

Kizlink, Juraj. 2014. *Odpady: sběr, zpracování, zužitkování, zneškodnění, legislativa* [Waste: Collection, Processing, Use, Disposal, Legislation]. Brno: CERM.

Kjeldsen, Peter, Morton A Barlaz, Alix P Rooker, Anders Baun, Anna Ledin, and Thomas H Christensen. 2002. "Present and Long-Term Composition of MSW Landfill Leachate: A Review." *Critical Reviews in Environmental Science and Technology* 32(4): 297–336.

Kliková, Jana, and Lenka Brunclíková. 2017. "Dumpster Diving: Mezi etikou a zákonem." *Antropowebzin* 2017(1–2): 31–37.

Klíma, Ivan. 1991[1988]. *Love and Garbage.* New York: Alfred A. Knopf.

Knight, Daniel M. 2021. *Vertiginous Life: An Anthropology of Time and the Unforeseen.* Oxford: Berghahn Books.

Knox, Keith, Richard P. Beaven, and Raffaello Cossu. 2018. "Chapter 12.1—Leachate Recirculation: History, Objectives, and Conceptual Design." In *Solid Waste Landfilling*, ed. Raffaello Cossu and Rainer Stegmann, 691–701. Amsterdam: Elsevier.

Knudsen, Ida H. and Martin D. Frederiksen. 2015. *Ethnographies of Grey Zones in Eastern Europe: Relations, Borders and Invisibilities.* London: Anthem Press.

Kohn, Eduardo. 2013. *How Forests Think: Toward an Anthropology Beyond the Human.* Berkeley: University of California Press.

Kojanić, Ognjen. 2020. "Theory from the Peripheries." *Anthropological Journal of European Cultures* 29(2): 49–66.

Kormaňáková, Marcela, Marcela Remešová, and Terézia Vančová. 2021. "Food Waste in Municipal Mixed Waste Produced at Household Level: Empirical Evidence from the Czech Republic." *Journal of Material Cycles and Waste Management* 23: 1348–64.

Kostkan, Vlastimil, Jana Laciná, Monika Mazalová, Petr Hekera, Vít Voženílek, and Jan Heisig. 2010. "Struktura směsného komunálního odpadu v různých typech zástavby v olomoucké aglomeraci." *Sborník konference Odpadové fórum 2010.* Kouty nad Desnou.

Krause, Franz, and Veronica Strang. 2016. Thinking Relationships Through Water. *Society & Natural Resources* 29(6): 633–38.

Kristeva, Julia. 1982. *Powers of Horror.* New York: Columbia University Press.

Křížová, Markéta. 2025. "Neither Red Enough nor Fierce Enough: The Construction of Native Americans in Nineteenth-Century Czech Culture." In *Habsburg Encounters with Native America: Subtitle: Familiar Strangers*, ed. Jonathan Singerton, Markéta Křížová, and Michael Burri, 111–34. Budapest: CEU Press.

Kubíčková, Lea, Lucie Veselá, and Marcela Kormaňáková. 2021. "Food Waste Behaviour at the Consumer Level: Pilot Study on Czech Private Households." *Sustainability* 13(20): 11311.

Kuna, Martin. 2015. "Categories of Settlement Discard." In *Paradigm Found: Archaeological Theory Present, Past and Future. Essays in Honour of Evžen Neustupný*, ed. Kristian Kristiansen, Ladislav Šmejda and Jan Turek, 278–92. Oxford: Oxbow.

Kuřík, Bob. 2021. "Kalistenika znovuobjevování s Jamesem C. Scottem." In *Dvakrát sláva anarchismu, šest prostých textů o autonomii, důstojnosti, smysluplné práci a hře*, ed. James C. Scott, 219–78. Praha: Svobodné knihovny.

Kürti, László. 2008. "East and West: The Scholarly Divide in Anthropology." *Anthropological Notebooks* 14(3): 25–38.

Laidlaw, James. 2010. "Agency and Responsibility: Perhaps You Can Have Too Much of a Good Thing." In *Ordinary Ethics: Anthropology, Language, and Action*, ed. Michael Lambek, 143–64. New York: Fordham University Press.

Lambek, Michael. 2010a. "Introduction." In *Ordinary Ethics: Anthropology, Language, and Action*, ed. Michael Lambek, 1–36. New York: Fordham University Press.

Lambek, Michael. 2010b. *Ordinary Ethics: Anthropology, Language, and Action.* New York: Fordham University Press.

Lambek, Michael. 2010c. "Toward an Ethics of the Act." In *Ordinary Ethics: Anthropology, Language, and Action*, ed. Micheal Lambek, 39–63. New York: Fordham University Press.

Laporte, Dominique. 2000[1978]. *History of Shit.* Cambridge: MIT Press.

Ledeneva, Alena. 1998. *Russia's Economy of Favours: Blat, Networking and Informal Exchange.* Cambridge: Cambridge University Press.

Ledeneva, Alena. 2011. "Open Secrets and Knowing Smiles." *East European Politics & Societies* 25(4): 720–36.

Ledeneva, Alena. 2018. "Introduction: The Informal View of the World: Key Challenges and Main Findings of the Global Informality Project." In *The Global Encyclopaedia of Informality, Volume 1*, ed. Alena Ledeneva, Anna Bailey, Sheelagh Barron, Costanza Curro, and Elizabeth Teague, 1–28. London: UCL Press.

Ledeneva, Alena, Anna Bailey, Sheelagh Barron, Costanza Curro, and Elizabeth Teague, eds. 2018. *The Global Encyclopaedia of Informality: Understanding Social and Cultural Complexity*. London: UCL Press.

Lednická, Karin. 2020. *Šikmý kostel: románová kronika ztraceného města, léta 1894–1921* [The Leaning Church: Chronicle of a Town that vanished, 1894–1921]. Ostrava: Bílá vrána.

Lepawsky, Josh. 2018. *Reassembling Rubbish: Worlding Electronic Waste.* Cambridge: The MIT Press.

Lévi-Strauss, Claude. 1966[1962]. *The Savage Mind.* Chicago: University of Chicago Press.

Liboiron, Max. 2009. "Recycling as a Crisis of Meaning." *eTopia.* Retrieved 20 September 2025 from https://discardstudies.com/2012/09/20/recycling-as-a-crisis-of-meaning/.

Liboiron, Max. 2019. "Waste Is Not 'Matter Out of Place'." *Discard Studies 09/09/2019.* Retrieved 20 March 2021 from https://discardstudies.com/2019/09/09/waste-is-not-matter-out-of-place/.

Liboiron, Max. 2021. *Pollution is Colonialism.* Durham: Duke University Press.

Liboiron, Max. 2022. "Matter Out of Place." In *The Routledge Handbook of Waste Studies*, ed. Zsusza Gille and Josh Lepawsky, 31–40. London: Routledge.

Liboiron, Max, and Josh Lepawsky. 2022. *Discard Studies: Wasting, Systems, and Power.* Cambridge: The MIT Press.

Liboiron, Max, Manuel Tironi, and Nerea Calvillo. 2018. "Toxic Politics: Acting in a Permanently Polluted World." *Social Studies Science* 48(3): 331–49.

Lien, Marianne E. 2015. *Becoming Salmon: Aquaculture and the Domestication of a Fish.* Oakland: University of California Press.

Lock, Margaret. 1996. "Displacing Suffering: The Reconstruction of Death in North America and Japan." *Daedalus* 125(1): 207–44.

Lozoviuk, Petr. 2012. *Grenzland als Lebenswelt: Grenzkonstruktionen, Grenzwahrnehmungen und Grenzdiskurse in sächsischtschechischer Perspektive.* Leipzig: Leipziger Universitätsverlag.

Lupták Burzová, Petra. 2019. "Být naplánován / Geografie 'nekvalifikovaných dělníků a dělnic'." In *Geografie 'okrajem:' Každodenní časoprostorové zkušenosti*, ed. Robert Osman and Lucie Pospíšilová, 91–110. Praha: Karolinum.

Ma, Shijun, Chuanbin Zhou, Jingjin Pan, Guang Yang, Chuanlian Sun, Yijie Liu, Xinchuang Chen, and Zhilan Zhao. 2022. "Leachate from Municipal Solid Waste Landfills in a Global Perspective: Characteristics, Influential Factors and Environmental Risks." *Journal of Cleaner Production* 333: 130234.

MacBride, Samantha. 2011. *Recycling reconsidered: the present failure and Future Promise of Environmental Action in the United States.* Cambridge: The MIT Press.

Makovicky, Nicolette. 2018. "Kombinowanie: agency, informality, and the poetics of self in Highland Poland." *Journal of the Royal Anthropological Institute* 24(3): 493–511.

Mandel, Ruth, and Caroline Humphrey, eds. 2002. *Markets and Moralities: Ethnographies of Post-Socialism.* Oxford: Berg.

Marcus, George E. 1986. "Contemporary Problems of Ethnography in the Modern World System." In *Writing culture: The Poetics and Politics of Ethnography: A School of American Research Advanced Seminar*, ed. James Clifford and George E. Marcus, 165–193. Berkeley: University of California Press.

Martínez, Francisco. 2019. "Insiders' Manual to Breakdown." In *Repair, Brokenness, Breakthrough: Ethnographic Responses*, ed. Francisco Martínez and Patrick Laviolette, 1–16. New York: Berghahn.

Martínez, Francisco, and Kaia Beilmann. 2020. "Waste and Postsocialism in Estonia: Becoming European Through the Management of Rubbish." *Environment and Planning C: Politics and Space* 38(7–8): 1348–66.

Martínez, Francisco. 2025. *The Future of Hiding: Secrecy, Infrastructure, and Ecological Memory in Estonia's Siberia.* Ithaca: Cornell University Press.

Martínez, Francisco, and Patrick Laviolette. 2019. *Repair, Brokenness, Breakthrough: Ethnographic Responses.* New York: Berghahn.

Marx, Karl. 1990[1867]. *Capital: A Critique of Political Economy.* Vol. 1. London: Penguin Books.

Mašek, Pavel. 2022. "Grey Greening: Quiet Sustainability at Auto Salvage Yards." *Geografie* 127(1): 55–73.

Mašek, Pavel. 2023. "'It's Only Us, Hyenas, Who Profit Out of It': Wrecked Cars, Leaked Humans, and the Death of the Person-car." *Journal of Material Culture* 28(1): 24–39.

McDougall, Debra. 2021. "Trash and Treasure: Pathologies of Permanence on the Margins of Our Plastic Age." In *Decay*, ed. Ghassan Hage, 28–36. Durham: Duke University Press.

McDougall, Forbes R., Peter R. White, Marina Franke, and Peter Hindle. 2008. *Integrated Solid Waste Management: A Life Cycle Inventory.* Oxford: Blackwell.

Melosi, Martin V. 2005. *Garbage in the Cities: Refuse Reform and the Environment.* Pittsburgh: University of Pittsburgh Press.

Melosi, Martin V. 2016. "Fresh Kills: The Making and Unmaking of a Wastescape." *RCC Perspectives* (1): 59–66.

Melosi, Martin V. 2020. *Fresh Kills: A History of Consuming and Discarding in New York City*. New York: Columbia University Press.

Metcalf, Peter, and Richard Huntington. 1991. *Celebrations of Death: The Anthropology of Mortuary Ritual.* Cambridge: Cambridge University Press.

Millar, Kathleen M. 2008. "Making Trash into Treasure: Struggles for Autonomy on a Brazilian Garbage Dump." *Anthropology of Work Review* 29(2): 25–34.

Millar, Kathleen M. 2012. "Trash Ties: Urban Politics, Economic Crisis and Rio de Janeiro's Garbage Dump." In *Economies of Recycling,* ed. Catherine Alexander and Joshue Reno, 164–84. London: Zed Books.

Millar, Kathleen M. 2018. *Reclaiming the Discarded: Life and Labor on Rio's Garbage Dump*. Durham: Duke University Press.

Millar, Kathleen M. 2020. "Garbage as Racialization." *Anthropology and Humanism* 45(1): 4–24.

Millar, Neal, Eoin McLaughlin, and Tobias Börger. 2019. "The Circular Economy: Swings and Roundabouts?" *Ecological Economics* 158: 11–19.

Miller, Daniel. 2001. "Alienable Gifts and Inalienable Commodities." In *The Empire of Things: Regimes of Value and Material Culture,* ed. Fred R. Myers, 91–115. Santa Fe: School of American Research.

Mol, Annemarie. 2002. *The Body Multiple: Ontology in Medical Practice*. Durham: Duke University Press.

Moreau, Vincent, Marlyne Sahakian, Pascal Van Griethuysen, and François Vuille. 2017. "Coming Full Circle: Why Social and Institutional Dimensions Matter for the Circular Economy." *Journal of Industrial Ecology* 21(3): 497–506.

Morris, Jeremy, and Abel Polese. 2014. *The Informal Post-Socialist Economy: Embedded Practices and Livelihoods*. London: Routledge.

Muller, Jerry Z. 2019. *The Tyranny of Metrics*. Princeton: Princeton University Press.

Müller, M. 2021. "Worlding Geography: From Linguistic Privilege to Decolonial Anywheres." *Progress in Human Geography* 45(6): 1440–66.

Nagle, Robin. 2013. *Picking Up: On the Streets and Behind the Trucks with the Sanitation Workers of New York City*. New York: Farrar, Straus and Giroux.

Nas, Peter J. M., and Rivke Jaffe. 2004. "Informal Waste Management." *Environment, Development and Sustainability* 6(3): 337–53.

Nešporová, Olga. 2021. *Funerary Practices in the Czech Republic*. Bingley: Emerald Publishing.

Niskanen, Johan, Jonas Anshelm, and Duncan McLare. 2020. "Local Conflicts and National Consensus: The Strange Case of Circular Economy in Sweden." *Journal of Cleaner Production* 261: 1–9.

Nixon, Rob. 2011. *Slow Violence and the Environmentalism of the Poor*. Cambridge: Harvard University Press.

Nordstrom, Carolyne. 2007. *Global Outlaws: Crime, Money, and Power in the Contemporary World.* Berkeley: University of California Press.

O'Hare, Patrick. 2018. "'The Landfill has Always Borne Fruit': Precarity, Formalisation and Dispossession Among Uruguay's Waste Pickers." *Dialectical Anthropology* 43(1): 31–44.

O'Hare, Patrick. 2020. "'We Looked After People Better when We Were Informal': The 'Quasi-Formalisation'of Montevideo's Waste-Pickers." *Bulletin of Latin American Research* 39(1): 53–68.

O'Hare, Patrick. 2021. "Cambridge, Carnaval, and the 'Actually Existing Circularity' of Plastics." *Worldwide Waste: Journal of Interdisciplinary Studies* 4(1): 1–12.

O'Hare, Patrick. 2022a. *Rubbish Belongs to the Poor: Hygienic Enclosure and the Waste Commons.* London: Pluto Press.

O'Hare, Patrick. 2022b. "Landfill Life and the Many Lives of Landfills." In *The Routledge Handbook of Waste Studies*, ed. Zsusza Gille and Josh Lepawsky, 55–67. London: Routledge.

O'Hare, Patrick, and Dagna Rams. 2024a. *Circular Economies in an Unequal World: Waste, Renewal and the Effects of Global Circularity.* New York: Bloomsbury.

O'Hare, Patrick, and Dagna Rams. 2024b. "Introduction – Circular Economies: Between the Promise of Renewal and Unequal Global Circulation." In *Circular Economies in an Unequal World: Waste, Renewal and the Effects of Global Circularity*, ed. Patrick O'Hare and Dagna Rams, 1–21. New York: Bloomsbury.

Okely, Judith. 1983. *The Traveller-Gypsies.* Cambridge: Cambridge University Press.

Okely, Judith. 1996. *Own or Other Culture.* London: Routledge.

Olsen, Bjørnar. 2010. *In Defense of Things: Archaeology and the Ontology of Objects.* Lanham: Rowman & Littlefield.

Olsen, Bjørnar, Michael Shanks, Timothy Webmoor, and Christopher Witmore. 2012. *Archaeology: The Discipline of Things.* Berkeley: University of California Press.

Ort, Jan. 2022. *Facets of a Harmony: The Roma and Their Locatedness in Eastern Slovakia.* Prague: Karolinum Press.

Ortner, Sherry B. 1995. "Resistance and the Problem of Ethnographic Refusal." *Comparative Studies in Society and History* 37(1): 173–93.

Packard, Vance. 1960. *The Waste Makers.* New York: Van Rees Press.

Parker Pearson, Michael. 1982. "Mortuary Practices, Society and Ideology: An Ethnoarchaeological Study." In *Symbolic and Structural Archaeology*, ed. Ian Hodder, 99–113. Cambridge: Cambridge University Press.

Parry, Jonathan, and Maurice Bloch. 1989. "Introduction: Money and the Morality of Exchange." In *Money and the Morality of Exchange*, Jonathan Parry and Maurice Bloch, 1–32. Cambridge: Cambridge University Press.

Pauknerová, Karolína. 2019. *Krajina mezi pamětí a zapomínáním* [The Landscape between Memory and Forgetting]. Praha: Karolinum.

Pavlínek, Petr. 2014. "Whose Success? The State–Foreign Capital Nexus and the Development of the Automotive Industry in Slovakia." *European Urban and Regional Studies* 23(4): 571–93.

Pavlínek, Petr, and John Pickles. 2002. *Environmental Transitions: Transformation and Ecological Defense in Central and Eastern Europe.* London: Routledge.

Pawlik, Wojciech. 1992. "Intimate Commerce." In *The Unplanned Society: Poland During and After Communism*, ed. Janine R. Wedel, 78–94. New York: Columbia University Press.

Pehe, Veronika. 2020. *Velvet Retro: Postsocialist Nostalgia and the Politics of Heroism in Czech Popular Culture.* New York: Berghahn Books.

Pelikán, Vojtěch. 2010. "Vztah spišských Romů k přírodě ve světle antropologických teorií." *Antropowebzin* 3(2010): 205–10.

Pétursdóttir, Þóra, and Bjørnar Olsen. 2014. "An Archaeology of Ruins." In *Ruin Memories: Materialities, Aesthetics and the Archaeology of the Recent Past*, ed. Bjørnar. Olsen and Þóra. Pétursdóttir, 1–29. New York: Routledge.

Piasere, Leonardo. 1985. *Mare Roma, catégories humaines et structure sociale. Une contribution à l'ethnologie tsigane.* Paris: Paul H. Stahl.

Pichtel, John. 2014. *Waste Management Practices: Hazardous, Municipal, and Industrial.* Boca Raton: Taylor & Francis.

Podjed, Dan, and Katarina Polajnar Horvat. 2020. "The Invisible Life of Food Waste." *Traditiones* 49(1): 109–24.

Polansky, Paul. 2014. *Tábor smrti Lety: vyšetřování začíná: 1992–1995* [Death Camp Lety: The Investigation Begins (1992-1995)]. Antifašistická akce.

Polanyi, Karl. 1944. *The Great Transformation.* New York: Farrar & Rinehart.

Polese, Abel. 2023. "What is Informality? (Mapping) 'The Art of Bypassing the State' in Eurasian Spaces – and Beyond." *Eurasian Geography and Economics* 64(3): 1–43.

Polese, Abel, and Jeremy Morris. 2015. "My Name is Legion. The Resilience and Endurance of Informality Beyond, or in Spite of, the State." In *Informal Economies in Post-Socialist Spaces – Practices, Institutions and Networks,* ed. Jeremy Morris and Abel Polese, 1–21. London: Palgrave Macmillan.

Pospíšil, Leopold J. 2021. *Adventures in the Stone Age: A New Guinea Diary.* Praha: Karolinum.

Pulay, Gergely. 2015. "The Street Economy in a Poor Neighbourhood of Bucharest." In *Gypsy Economy: Romani Livelihoods and Notions of Worth in the 21st Century,* ed. Micol Brazzabeni, Manuela Ivone Pereira da Cunha, and Martin Fotta, 127–44. New York: Berghahn.

Pulay, Gergely. 2023. "Personalized Value Struggles amid Marketization: The Search for the Good Among Men on the Margins of Bucharest." *East European Politics and Societies* 38(1): 339–56.

Rakowski, Tomasz. 2016. *Hunters, Gatherers, and Practitioners of Powerlessness: An Ethnography of the Degraded in Postsocialist Poland.* New York: Berghahn Books.

Rakowski, Tomasz. 2017. "Interior Spectacles: The Art of the Informal Among Bootleg Miners in Wałbrzych, Poland." In *Economies of Favour After Socialism,* ed. David Henig and Nicolette Makovicky, 161–80. Oxford: Oxford University Press.

Rathje, William L., and Cullen Murphy. 2001. *Rubbish!: The Archaeology of Garbage.* Tucson: University of Arizona Press.

Reno, Joshua O. 2009. "Your Trash Is Someone's Treasure: The Politics of Value at a Michigan Landfill." *Journal of Material Culture* 14(1): 29–46.

Reno, Joshua O. 2014. "Toward a New Theory of Waste: From 'Matter out of Place' to Signs of Life." *Theory, Culture & Society* 31(6): 3–27.

Reno, Joshua O. 2015. "Waste and Waste Management." *Annual Review of Anthropology* 44: 557–72.

Reno, Joshua O. 2016a. *Waste Away: Working and Living with a North American Landfill.* Oakland: University of California Press.

Reno, Joshua O. 2016b. "The Life and Times of Landfills." *Journal of Ecological Anthropology* 18. Retrieved 11 January 2017 from https://scholarcommons.usf.edu/jea/vol18/iss1/15.

Reno, Joshua O. 2018. "Kept in Suspense: The Unsettling Indeterminacy of US Landfills." In *Indeterminacy: Waste, Value, and the Imagination,* ed. Catherine Alexander and Andrew Sanchez, 31–49. New York: Berghahn.

Reno, Joshua O. 2020. *Military Waste: The Unexpected Consequences of Permanent War Readiness.* Oakland: University of California Press.

Resnick, Elana. 2018. "Durable Remains: Glass Reuse, Material Citizenship and Precarity in EU-era Bulgaria." *Journal of Contemporary Archaeology* 5(1): 103–15.

Resnick, Elana. 2021. "The Limits of Resilience: Managing Waste in the Racialized Anthropocene." *American Anthropologist* 123(2): 222–36.

Resnick, Elana. 2024a. "The Determined Indeterminacy of White Supremacy." *American Ethnologist* 51(3): 433–47.

Resnick, Elana. 2024b. "Sustaining Containability: Zero Waste and White Space." *Cultural Anthropology* 39(2): 216–45.

Resnick, Elana. 2025. *Refusing Sustainability: Race and Environmentalism in a Changing Europe*. Stanford: Stanford University Press.

Robben, Antonius C. G. M. 1995. "The Politics of Truth and Emotion among Victims and Perpetrators of Violence." In *Fieldwork under Fire: Contemporary Studies of Violence and Culture*, ed. Carolyn Nordstrom and Antonius C. G. M. Robben, 81–103. Berkeley: University of California Press.

Robben, Antonius C. G. M. 2004a. *Death, Mourning, and Burial: A Cross-Cultural Reader.* Malden: Blackwell.

Robben, Antonius C. G. M. 2004b. "State Terror in the Netherworld: Disappearance and Reburial in Argentina." In *Death, Mourning, and Burial: A Cross-Cultural Reader*, Antonius C. G. M. Robben, 134–48. Malden: Blackwell.

Robben, Antonius C. G. M. 2018. *Argentina Betrayed: Memory, Mourning, and Accountability.* Philadelphia: University of Pennsylvania Press.

Rubbers, Benjamin. 2007. "Retour sur le 'secteur informel': L'économie du Katanga (Congo-Zaïre) face à la falsification de la loi." *Sociologie Du Travail* 49(3): 316–29.

Saad, Hwaida, and Anne Barnard. 2015. "Government and Protesters in Lebanon Weigh Next Move in Trash Dispute." *The New York Times* Aug. 24, 2015. Retrieved 23 March 2016 from https://www.nytimes.com/2015/08/25/world/middleeast/protests-in-beirut-over-uncollected-garbage-turn-violent.html.

Sacthre, Eirik. 2020. *Wastelands: Recycled Commodities and the Perpetual Displacement of Ashkali and Romani Scavengers.* Oakland: University of California Press.

Sahlins, Marshall. D. 1972. *Stone Age Economics.* Chicago: Aldine-Atherton.

Sahlins, Marshall. D. 1976a. *Culture and Practical Reason.* Chicago: University of Chicago Press.

Sahlins, Marshall. D. 1976b. *The Use and Abuse of Biology: An Anthropological Critique of Sociobiology.* Ann Arbor: University of Michigan Press.

Sanchez, Andrew. 2016. *Criminal Capital: Violence, Corruption and Class in Industrial India*, New Delhi: Routledge.

Sanchez, Andrew. 2020. "Transformation and the Satisfaction of Work." *Social Analysis* 64(3): 68–94.

Schiffer, Michael B. 1996. *Formation Processes of the Archaeological Record.* Salt Lake City: University of Utah Press.

Schling, Hannah. 2017. "(Re)Production: Everyday Life in the Workers' Dormitory." *Society and Space Magazine*, Nov 7.

Schmidt, Michelle. 2022. "Commodification and Respect: Indigenous Contributions to the Sociology of Waste." In *The Routledge Handbook of Waste Studies*, ed. Zsusza Gille and Josh Lepawsky, 103–18. London: Routledge.

Scott, James C. 1985. *Weapons of The Weak: Everyday Forms of Peasant Resistance.* New Haven: Yale University Press.

Serres, Michel. 2011. *Malfeasance: Appropriation Through Pollution?* Stanford: Stanford University Press.

Šiklová, Jiřina, Káča Poláčková-Henley, and Gerald Turner. 1990. "The 'Gray Zone' and the Future of Dissent in Czechoslovakia. *Social Research* 57(2): 347–63.

Simmel, Georg, and Kurt H. Wolff. 1950. *The Sociology of Georg Simmel.* Glencoe: Free Press.

Skalník, Petr. 2018. "Introduction: Opole Power Station as a Socio-Political Catalyzer." In *Mutual Impact: Conflict, Tension and Cooperation in Opole Silesia*, ed. Petr Skalník, 19–43. Wroclaw: University of Wroclaw.

Škobla, Daniel, and Richard Filčák. 2023. "Life Next to a Landfill: Urban Marginality, Environmental Injustice and the Roma." *Race & Class* 65(4): 74–91.

Smith, Joe, and Petr Jehlička. 2013. "Quiet Sustainability: Fertile Lessons from Europe's Productive Gardeners." *Journal of Rural Studies* 32: 148–57.

Solimene, Marco. 2015. "'I Go for Iron': Xoraxané Romá Collecting Scrap Metal in Rome." In *Gypsy Economy: Romani Livelihoods and Notions of Worth in the 21st Century*, ed. Micol Brazzabeni, Manuela Ivone Pereira da Cunha, and Martin Fotta, 107–26. New York: Berghahn.

Sørensen, Tim F. 2010. "A Saturated Void: Anticipating and Preparing Presence in Contemporary Danish Cemetery Culture." In *An Anthropology of Absence: Materializations of Transcendence and Loss*, ed. Mikkel Bille, Frida Hastrup, and Tim Flohr Sørensen, 115–30. New York: Springer.

Sosna, Daniel. 2017. "Heterotopias Behind the Fence: Landfills as Relational Emplacements." In *Archaeologies of Waste: Encounters with the Unwanted*, ed. Daniel Sosna and Lenka Brunclíková, 162–78. Oxford: Oxbow Books.

Sosna, Daniel. 2022. "Saving and Wasting: The Paradox of Thrift in a Czech Landfill." In *Thrift and Its Paradoxes: From Domestic to Political Economy*, ed. Catherine Alexander and Daniel Sosna, 162–84. New York: Berghahn.

Sosna, Daniel. 2023. "Magnetism of strangeness: Silenced histories of landscapes." *History and Anthropology* 34(3): 390–408.

Sosna, Daniel. 2024. "The Inner Dynamics of Moral Economies: The Case of Waste Management." *East European Politics and Societies* 38(1): 321–38.

Sosna, Daniel. 2024. "Circular Economy of Wastewater: Recirculation, Spinning, and Rolling to the Future." In *Circular Economies in an Unequal World: Waste, Renewal, and the Effects of Global Circularity*, ed. Patrick O'Hare and Dagna Rams, 153–71. New York: Bloomsbury.

Sosna, Daniel, and Lenka Brunclíková. 2015. *Odpad pohledem společenských věd: metodická příručka* [Waste from a Social Science Perspective: A Handbook of Methods]. Plzeň: Západočeská univerzita.

Sosna, Daniel, and Lenka Brunclíková, eds. 2017. *Archaeologies of Waste: Encounters with the Unwanted.* Oxford: Oxbow Books.

Sosna, Daniel, and Lenka Brunclíková. 2019. "Drinking and Dreaming in the Dormitories: Workers Struggle in Times of Economic Development." *Journal of Material Culture* 24(2): 232–50.

Sosna, Daniel, Lenka Brunclíková, and Patrik Galeta. 2019. "Rescuing Things: Food Waste in the Rural Environment in the Czech Republic." *Journal of Cleaner Production* 214: 319–30.

Sosna, Daniel, Barbora Stehlíková, and Pavel Mašek. 2024. "Ecologies of Quantification in Waste Management: Landfilling, e-Waste Recycling, and Car Breaking." *Critique of Anthropology* 44(1): 42–63.

Sovová, Lucie, Petr Jehlička, and Petr Daněk. 2021. "Growing the Beautiful Anthropocene: Ethics of Care in East European Food Gardens." *Sustainability* 13(9): 5193.

Stamatopoulou-Robbins, Sophia. 2020. *Waste Siege: The Life of Infrastructure in Palestine.* Stanford: Stanford University Press.

Stan, Sabina. 2012. "Neither Commodities nor Gifts: Post-Socialist Informal Exchanges in the Romanian Healthcare System." *Journal of the Royal Anthropological Institute* 18(1): 65–82.

Stegmann, Rainer. 2018. "10.1—Strategic Issues in Leachate Management." In *Solid Waste Landfilling*, ed. Raffaello Cossu and Rainer Stegmann, 501–509. Amsterdam: Elsevier.

Stehlíková, Barbora. 2021. "'The Aim is Always Joy!' The Hybrid Gift as an Anchor for Morality." *Cargo*(1–2): 86–104.

Stehlíková, Barbora. 2024. "E-waste between Morality and Ethics: Waste Practices in the Czech Republic." PhD dissertation, Department of Sociology, Institute of Sociological Studies, Faculty of Social Sciences, Charles University, Prague.

Stehlíková, Barbora. 2025. "Classification Freedom." *Anthropological Journal of European Cultures* 34(1): 103–22.

Stewart, Charles. 2017. "Uncanny History: Temporal Topology in the Post-Ottoman World." *Social Analysis* 61(1): 129–42.

Stewart, Michael. 1997. *The Time of the Gypsies: Studies in the Ethnographic Imagination.* Boulder: Westview Press.

Stowell, Alison F., and Samantha Warren. 2018. "The Institutionalization of Suffering: Embodied Inhabitation and the Maintenance of Health and Safety in E-waste Recycling." *Organization Studies* 39(5–6): 785–809.

Strain, Daniel. 2012. "Ocean Trash Is a Lifesaver for Insect." *ScienceMag.org.* American Association for the Advancement of Science. Retrieved 23 September 2020 from www.science.org/content/article/ocean-trash-lifesaver-insect.

Strang, Veronica. 2014. "Fluid Consistencies. Material Relationality in Human Engagements with Water." *Archaeological Dialogues* 21(2): 133–50.

Strathern, Marilyn. 1992. "Parts and Wholes: Refiguring Relationships in a Post-Plural World." In *Conceptualizing Society*, ed. Adam Kuper, 75–104. London: Routledge.

Szczygieł, Mariusz. 2014[2007]. *Gottland: Mostly True Stories from Half of Czechoslovakia.* New York: Melville House.

Taussig, Michael. T. 1980. *The Devil and Commodity Fetishism in South America.* Chapel Hill: University of North Carolina Press.

Teng, Chunying, Kanggen Zhou, Changhong Peng, and Wei Chen. 2021. "Characterization and Treatment of Landfill Leachate: A Review." *Water Research* 203: 117525.

Thelen, Tatjana. 2011. "Shortage, Fuzzy Property and Other Dead Ends in the Anthropological Analysis of (Post)Socialism." *Critique of Anthropology* 31(1): 43–61.

Thiemann, André. 2024. "Infrastructuring 'Red Gold': Agronomists, Cold Chains, and the Involution of Serbia's Raspberry Country." *Ethnos* 89(2): 289–311.

Tremčinský, Martin. 2022. "Bitcoin and its Spheres of Consumption: Transactional Orders of Consuming Money in the Czech and Slovak Bitcoin Community." *Economic Anthropology* 9(1): 35–46.

Trouillot, Michel-Rolph. 2003. *Global Transformations: Anthropology and the Modern World.* New York: Palgrave Macmillan.

Tsing, Anna L. 2005. *Friction: An Ethnography of Global Connection.* Princeton: Princeton University Press.

Tsing, Anna L. 2015. *The Mushroom at the End of the World: On the Possibility of Life in Capitalist Ruins.* Princeton: Princeton University Press.

Tsing, Anna L., Heather A. Swanson, Elaine Gan, and Nils Bubandt. 2017. *Arts of Living on a Damaged Planet.* Minneapolis: University of Minnesota Press.

Turner, Victor. 1967. *The Forest of Symbols: Aspects of Ndembu Ritual.* Ithaca: Cornell University Press.

Ucko, Peter J. 1969. "Ethnography and Archaeological Interpretation of Funerary Remains." *World Archaeology* 1: 262–80.

Uhlová, Saša. 2018. *Hrdinové kapitalistické práce* [Heroes of Capitalist Work]. Praha: Grada Publishing.

Vašát, Petr. 2014. "'Předevčírem, nebo kdy to bylo?': Temporalita třídy nejchudších." *Sociologický časopis* 50(1): 57–82.

Vašát, Petr. 2023. "Quasi-Household Economy: Rethinking Homelessness in Post-Socialist Pilsen and Beyond." *Cities* 134: 104190.

Vaverková, Magdalena D. 2019. "Landfill Impacts on the Environment - Review." *Geosciences* 9(10): 431–47.

Vávra, Jan, Petr Daněk, and Petr Jehlička. 2018. "What is the Contribution of Food Self-Provisioning Towards Environmental Sustainability? A Case Study of Active Gardeners." *Journal of Cleaner Production* 185: 1015–23.

Verdery, Katherine. 1999. *The Political Lives of Dead Bodies: Reburial and Postsocialist Change.* New York: Columbia University Press.

Vergara, Sintana E., and George Tchobanoglous. 2012. "Municipal Solid Waste and the Environment: A Global Perspective." *Annual Review of Environment and Resources* 37(1): 277–309.

Veselá, Lucie, Martina Vršanská, Irena Baláková, Stanislava Voběrková, Lea Kubíčková, and Magdalena D. Vaverková. 2025. "Analysis of the Structure and Seasonal Variability of Food Waste in Urbanized Areas: Possible Actions for Achieve Sustainable Development." *Sustainable Development* 33(4): 5742–61.

Wanner, Catherine. 2005. "Money, Morality and New Forms of Exchange in Postsocialist Ukraine." *Ethnos* 70(4): 515–37.

Weber, Heike. 2022a. "Recycling, Remaking and Reusing in 20th-Century Europe: From a Culture of Thrift to One of Abundance." In *Throwaway: The History of Modern Crisis*, ed. Christine Dupont, Stéphanie Gonçalves and Emma Teworte, 94–107. Luxembourg: Publications Office of the European Union.

Weber, Heike. 2022b. "Unmaking the Made: The Troubled Temporalities of Waste." In *The Routledge Handbook of Waste Studies*, ed. Zsusza Gille and Josh Lepawsky, 88–102. London: Routledge.

Wedel, Janine. 1992. *The Unplanned Society: Poland During and After Communism.* New York: Columbia University Press.

Weiner, Annette B. 1992. *Inalienable Possessions: The Paradox of Keeping-While-Giving.* Berkeley: University of California Press.

Widlok, Thomas. 2016. *Anthropology and the Economy of Sharing.* London: Taylor & Francis.

Woitsch, Jiří, and Karolína Pauknerová. 2019. "From the Intelligence Tower to the Tower of Joy: The Story of Havran in the Former Iron Curtain Buffer Zone." *Český lid* 106(4): 409–37.

Wolff, Larry. 1994. *Inventing Eastern Europe: The Map of Civilization on the Mind of the Enlightenment*. Stanford: Stanford University Press.

Wölfle Hazard, Cleo. 2022. *Underflows: Queer Trans Ecologies and River Justice.* Seattle: University of Washington Press.

Wynne, Brian. 1987. *Risk Management and Hazardous Waste: Implementation and the Dialectics of Credibility*. Berlin: Springer.

Xiezhi, Yu, Zennegg Markus, Engwall Magnus, Rotander Anna, Larrson Maria, Wong Ming Hung, and Weber Roland. 2008. "E-Waste Recycling Heavily Contaminates a Chinese City with Chlorinated, Brominated and Mixed Halogenated Dioxins." *Organohalogen Compounds* 70(1): 813–16.

Zani, Leah. 2019. *Bomb Children: Life in the Former Battlefields of Laos.* Durham: Duke University Press.

Zapata Campos, María J. 2013. "The Function of Waste Urban Infrastructures as Heterotopias of the City: Narratives from Gothenburg and Managua." In *Organising Waste in the City,* ed. María J. Zapata Campos and Michael Hall, 41–60. Bristol: Policy Press.

Zapata Campos, María J., and Patrik Zapata. 2012. "Changing La Chureca: Organizing City Resilience Through Action Nets." *Journal of Change Management* 12(3): 323–37.

Zapata, Patrik, and Zapata Campos, María J. 2015. "Producing, Appropriating and Recreating the Myth of the Urban Commons." In *Urban Commons: Rethinking the City*, ed. Christian Borch and Martin. Kornberger, 92–108. London: Routledge.

Zeitlyn, David. 2009. "Understanding Anthropological Understanding for a Merological Anthropology." *Anthropological Theory* 9(2): 209–31.

Zhang, Amy. 2020. "Circularity and Enclosures: Metabolizing Waste with the Black Soldier Fly." *Cultural Anthropology* 35(1): 74–103.

Zhang, Amy. 2024. *Circular Ecologies: Environmentalism and Waste Politics in Urban China*. Stanford: Stanford University Press.

Zhang, Chunbo, Mingming Hu, Francesco Di Maio, Benjamin Sprecher, Xining Yang, and Arnold Tukker. 2022. "An Overview of the Waste Hierarchy Framework for Analyzing the Circularity in Construction and Demolition Waste Management in Europe." *Science of The Total Environment* 803: 149892.

Zink, Robyn, and Lisette Burrows. 2008. "'Is What You See What You Get?' The Production of Knowledge In-Between the Indoors and the Outdoors in Outdoor Education." *Physical Education & Sport Pedagogy* 13(3): 251–65.

Žižek, Slavoj. 1997. *The Plague of Fantasies.* London: Verso.

INDEX

Numbers followed by 'n' refer to a numbered chapter endnote, e.g. 136n16.

www.ingramcontent.com/pod-product-compliance
Lightning Source LLC
LaVergne TN
LVHW011814200926
848484LV00005B/1189
9781836955511